Uncle Dave's Cow

LOCALLY
RAISED.

UNCLE DAVE'S COW

And Other Whole Animals
My Freezer Has Known

a guide
to sourcing,
storing, and
preparing
healthy, locally
raised meat

Leslie Miller

SKIPSTONE

Published by Skipstone, an imprint of The Mountaineers Books

Printed in the United States of America

First printing 2012
15 14 13 12   5 4 3 2 1

Copy Editor: Barry Foy
Design: Heidi Smets, heidismets.com
Layout: Jennifer Shontz, redshoedesign.com
Illustrations by Ayun Halliday

ISBN (paperback): 978-1-59485-697-6
ISBN (ebook): 978-1-59485-698-3

Library of Congress Cataloging-in-Publication Data

Miller, Leslie (Leslie Ann)
   Uncle Dave's cow and other whole animals my freezer has known: a guide to sourcing, storing, and preparing healthy, locally raised meat / Leslie Miller.
      pages   cm
   Includes index.
   ISBN 978-1-59485-697-6 (ppk.)—ISBN 978-1-59485-698-3 (ebook) (print)
   1. Slaughtering and slaughter-houses. 2. Meat—Storage. 3. Meat—Preservation.
4. Local foods. I. Title.
   TS1962.M55 2012
   664'.902—dc23
                                                                    2012025246

Skipstone books may be purchased for corporate, educational, or other promotional sales. For special discounts and information, contact our Sales Department at 800-553-4453 or mbooks@mountaineersbooks.org.

Skipstone
1001 SW Klickitat Way
Suite 201
Seattle, Washington 98134
206.223.6303
www.skipstonebooks.org
www.mountaineersbooks.org

SUSTAINABLE
FORESTRY
INITIATIVE
Certified Chain of Custody
Promoting Sustainable Forestry
www.sfiprogram.org
SFI-01268
SFI label applies to the text stock

LIVE LIFE. MAKE RIPPLES.

In loving memory of Dad,

who was always up for a good food adventure

CONTENTS

THE RECIPES

Cow

Pig

INTRODUCTION

This is a book about meat, about people who produce meat and people who eat it, and a consumer guide meant to change how you buy and consume meat in your daily life. This story traces a very personal journey that began the day my brother asked me to go in with him on buying a whole beef. Though my family lives a very urban lifestyle, both of my parents grew up on farms and plenty of my kin still earn their livelihoods through farming, including the titular Uncle Dave. As it turned out, being related to the farmer who raised my food encouraged a whole community to form surrounding this meat. As my brother and I cooked through the wrapped cuts in our freezers, we shared recipes, photos of finished dishes, and our experiences cooking the beef for our respective families.

Having begun with this quarter of a cow, I became intrigued by the idea of buying into different types of whole animals. I began to research breeds and growers and investigate price structures and ordering techniques. Soon I had gone halvsies on a local hog with my cousin Kate, and then buying meat this way, by the animal and not by the cut, became the new standard for my family. You will encounter more of Cousin Kate in the pages to come, as well as the rest of the Cast of Characters that starred in my personal production of meat and meals over the last couple of years and still do so today, from Uncle Dave, who raised that first cow, to my children (and dogs) who helped eat it.

In truth, buying a whole butchered animal or sharing in one for meat is not a new concept. My family did it when I was a girl. Like many contemporary urban obsessions, what's old is new again. When individual farmers slaughter individual animals, the result is a large amount of flesh, fat, and innards that need to be consumed or preserved to enjoy until the next slaughter. For the consumer, this is a time-honored way of getting meat at a good price, because you save both by buying in bulk and by buying a mix of cuts. Additionally, having meat on hand in the freezer makes it easy to produce daily meals without making a separate trip to the store.

However, what is different about my buying meat this way now is the conscious choices I—and you—can make about how that meat is produced. This book lays out those practices and the choices an average consumer is able to make when selecting a cow, pig, lamb, or goat to buy and eat. This book details for you the critical information needed to find a grower, plan for the right amount of meat, order the right cuts, pay for it, store it, and enjoy it. My goal in writing this book was also to explore and lay bare the myths and misconceptions *I* had at the outset, from small farms necessarily being better than larger ones to what it really feels like to know your food. I hope as you laugh at my naïveté, you are able to learn from my mistakes (as I have) and discover how easy it is to change how you buy and cook meat in potentially very important and delicious ways.

CAST OF CHARACTERS

» **Lam** Unrelated to the meat or the animal, I am author Leslie Ann Miller, and I am highly motivated by what I think of as "food fun" and adventure. A former vegetarian of twelve years, now a stalwart meat eater who cures sausages in her basement fitness room. Writer, editor, wife, and mother of two children in a hectic West Coast urban environment.

» **Uncle Dave** Youngest brother of my mother. Formerly in the dairy business, now a Western Washington farmer who raises a host of row crops as well as beef cows.

» **Cousin Kate** Oldest daughter of Uncle Dave. Wife and mother of three who raises pumpkins, cattle, and other crops as a large-scale "hobby." A rural lady of strong opinions and droll wit who finds "urban hippie back-to-the-landers," including Cousin Lam, mildly amusing.

» **Jason** Husband of Cousin Kate and father of three. Amiable veterinarian with a focus on large animals, who works in his practice with Western Washington farmers. Probably also finds his Cousin-in-Law Lam entertaining in her urban naïveté, but is too diplomatic to say so.

» **The Farm Children** Jackson, Margarite, and the Pterodactyl Baby, children of Cousin Kate and Jason. Find their urban second cousins' lifestyle—including not waking early for chores and lack of cattle knowledge—both intriguing and silly.

» **Jeff** Youngest of my three older brothers, dubbed a "guncle" or "gun uncle" by my children (this is a compliment). Formerly in the restaurant industry. Also motivated by "food fun" but more mindful of food costs and practicality than his sister. Conservative outdoorsman and hunter of game birds. A pivotal date with the woman who would become his wife included butchering pigs.

» **Erik** My middle brother. Foodie of more epicurean tastes, can still be coaxed into sausage-making sessions and suturing turduckens. The closest he comes to hunting is buying Italian leather shoes.

» **Dee** My oldest brother, cherished guncle and avid outdoorsman, fisherman, and hunter of birds, deer, and elk. Dentist by day. Longtime devotee of Grand Sausageman Rytek Kutas. Has a smoker jerry-rigged out of an old refrigerator in his backyard. Can be easily swayed into "food fun" participation.

» **Romeo** My youngest son. Seven-year-old who finds sauces suspicious but loves goat meatballs. Lactose intolerant and temporarily diagnosed with gluten and egg allergies as well. Enthusiastic endorser of the whole-animal lifestyle. Learned to read in part from retrieving meat packages from the downstairs freezer: "*Bay-kin*—oh, bacon!"

» **Atticus** My oldest son. Adventurous eater, save any meat that exhibits a shred of skin, fat, bone, tendon, or gristle. Seems to enjoy cheese even more since his brother is unable to eat it. Favorite whole-animal activity might be grinding meat through the KitchenAid attachment.

» **Husband Hendel** Left-brained software engineer who ranges from bemusedly supportive to completely put out by my perceived "food fun," especially by the sausage hanging above his skis. Even so, an enthusiastic sausage *eater*. Participation includes buying cookbooks and Cabela's food-processing gadgets for his wife. Got excited about building a curing chamber, though, once he learned it required a computer fan motor and hygrometer.

» **Remington and Robert Gregory** Bloodhound and black-and-tan coonhound, respectively. Large and naughty enough to collaboratively snatch a low-hanging skein of curing sausages. (Lack of consequent death did, however, provide evidence against the presence of botulism.) Eager endorsers of the whole-animal lifestyle, especially the cooking of bone-in meats that they might enjoy after the people are through.

1

• •

IT ALL BEGAN WITH UNCLE DAVE'S COW

My brother Jeff used to work in kitchens. He cooked, he managed. It made him a scrimper, a saver, someone interested in price points. This was his primary motivation for buying into one of Uncle Dave's cows. But for me, his phone call was all about adventure. Did I want to buy a quarter of a cow? Of course! I'd barely hung up the phone before I was shopping for an upright freezer.

I don't know what other people do with their brothers. Maybe they go to the movies or out for a beer. With mine, bonding has involved sausage marathons and holiday chowder-offs, homemade deer jerky and refrigerators-turned-smokers (more on that later). And now, bulk beef.

Jeff lived outside of Seattle at the time, much closer to the aunts and uncles and cousins that still live on the farm where my mother grew up in Western Washington. My childhood summers at Grams's house were replete with calves that licked your fingers with rough tongues and Charlie, the nearly mythical bull stationed out at the Tin Barn, who my brothers said would just as soon gore you as look at you. I remember the cows as relatively placid, spotted, and grand producers of milk.

Fast-forward a few decades, and the dairy business isn't what it used to be. My uncles and cousins have changed with the times, raising alfalfa, wheat, barley, sweet corn, peas, and other seed crops, to name just a few. Uncle

Dave now also keeps some beef cows, an Angus cross, raising them for meat instead of milk.

When Jeff called, the cow seemed like a good idea on many levels. I'm busy, my husband's busy, and my children have more active social lives than we do, and dinner isn't so much something to be crafted as it is a daily time-suck, with additional challenges posed by kids with rapidly cycling culinary preferences, and a husband who claims not to be picky but has stated, among other preferences, that he's "not a fan of discrete meat," meaning he'd rather his meat be well incorporated into a dish than centered on the plate with a starch to the side. Right. Throw in a liberal urban commitment to eating "good" meat and food in general, if possible—organic, sustainable, locally produced, all the buzzwords—and we seemed like good candidates for buying into that cow.

I wasn't grinding my own flour, but I was doing what I could, and a freezer full of varying cuts of beef seemed like it might help. Ground beef for cheeseburgers and meatballs and stew meat galore. Steaks—God, when was the last time I'd had a steak? And many large hunks of shoulder and rump that, in my fantasies, would cook low and slow with little effort and stretch into days of tasty leftovers that everyone would happily lap up.

On the big day, I loaded up the kids and a big cooler and made the drive down south. The cow had already been to the butcher, and what I encountered when I got to my brother's house was nothing more than a very large pile of tidy parcels wrapped in brown paper. RIBS, said one, ARM ROAST another. Stew meat, New York steaks, soup bones, ground beef. The packages were frozen solid and made satisfying noises as we tossed them into individual ice chests, trading cuts like baseball cards. Driving home, I was as proud as if I had a dressed deer tied to the top of my car that I'd downed with a bow and arrow. I felt it worth the gas guzzled by my four-wheel-drive to bring home over a hundred pounds of local grain- and grass-fed, humanely processed meat from an animal that was practically family.

Now, not everyone gets this family discount—I would find that out along my whole-animal journey—but my quarter-cow came at an incredible price (thank you, Uncle Dave). Comparatively, I felt I was paying mere pennies per pound for ground beef, and half of what I would pay for fancier cuts like prime rib and New York steaks at the local market. Not only did I have meals at the ready and have meat that was *good* in every sense of that word, but I was being thrifty at the same time.

Husband Hendel has long said that I am an Idea Person, an enthusiastic starter whose main source of enjoyment lies not in completing a task so much as in envisioning it. And three-quarters of the year into Uncle Dave's cow, I conceded that he may have had a point. There was still a lot of beef in that freezer. Don't get me wrong: We had been enjoying it. The flavor was incomparable, the color a deep, rich garnet. I loved having so much meat on hand, and we'd had delicious braises, roasts, steaks, burgers, meatballs, and more. But it also occurred to me—yes, after the fact—that my family doesn't really eat all that much beef. It didn't help that Atticus declared himself a vegetarian somewhere around that time, and then Husband Hendel went on a diet and asked for more meals centered around fish.

I feel like fish

Also, I finally did the math. A 1,200-pound cow translates into roughly 490 pounds of bone-in and boneless steaks and roasts, plus ground beef (a *lot* of ground beef). This means that my paltry quarter amounted to about 122 pounds of beef filling my freezer. If you ground all that up into quarter-pound burgers, do you know how many you would produce? Four hundred eighty-eight. When I was growing up, my mother would have run through that amount like a hot knife through butter, just feeding her four growing children. Cousin Kate still would, feeding her family of five.

I decided that while the cow had been a fun and delicious idea, maybe it wasn't the best choice for *our* family at the time. "A pig," I said to Husband Hendel. "We should buy a whole pig instead. Just think, we can cure our own bacon and prosciutto, and oh, the sausages we'll make!"

He was firm: "We have to get through the cow first." But as practical as he was, I could tell he was drawn to the pig. Even as I pulled out more ground beef to defrost, I schemed about pork. Unfortunately, Uncle Dave only raises cows, so a family connection wasn't going to fast-track me that bacon. But the more I thought about it, the more convinced I was that with a little research and, dare I say, *planning*, a pig could be mine. I did a little research…and I called my brothers to help me make sausages.

It has been two years since that first quarter-beef came to live with us. This book chronicles our journey from there onward through pig, goat, and lamb, all of which my freezer contains as I write this. To be honest, I started jotting down the anecdotes that would begin the book when I was struck by the rich irony in it all—my own lack of planning, Husband Hendel's rejection of meat-centered meals, the boys' insistent inconsistency on what they deemed edible from day to day. As I worked through the meat, got to know different growers and methods of animal agriculture, and got my info down, the process became less of a lark and more of a way of life.

The fun factor didn't totally dwindle. If you love to cook, this approach to meat opens up all sorts of new opportunities and experiences for you. Though I don't currently have an intact kitchen (due to an ill-timed remodel), I do have frozen sausages I made with my brothers, and the belly I just picked up with our latest half-pork was ordered uncured and uncut—I'll make my own bacon, thank you.

Even if making slow-cooker pot roast is your idea of fancy, I hope this book convinces you that, whether you're on the West Coast or in the Southeast, in a city or on a farm, ordering and eating whole animals, one at a time, is not just a good idea for my Cast of Characters, but for you, too.

cow eyeball (in cow)

MYTH #1

If I buy a whole animal, my children will never eat meat again!

It has been intriguing to observe, throughout this process of buying whole animals, that while adults purport to eschew whole-animal confrontation out of concern for the needs of their delicate youngsters, the young have no such hangups. This phenomenon was made clearest to me on the occasion of a recent birthday of mine (one that was a multiple of ten). Scores of adults and children went to a glorious campground on an island near Seattle, where we slept in old military barracks and played badminton and went to the beach and, on the second day, roasted an entire pig in my Caja China, a Cuban-style roasting box.

We had been referring to the pig as Cyrus III, he being the third whole or half-pig that I had procured and named Cyrus. But upon inspection of the body, it became clear that Cyrus had been of the female persuasion. She was a gilt, to use the precise term. As I loaded a syringe with *mojo de ajo* and jammed it here and there into her taut muscle, the adults did more than flinch. A couple rushed to get their four-year-old out of the room, a boy who was admiring her long face stretched back in a surprised grimace.

"Do you want to touch her?" I asked him, running my finger along her smooth pink skin. He followed suit, smiling as he brushed her skin and then poked his finger with more force into the muscle. I told him we should be thankful to this pig, for giving up her life so that we could enjoy her at the party. He gave her a pat, said thank you, and rushed off to play. Another three-year-old watched as Jason carefully disarticulated her head, our having discovered that she was too big to fit in the roasting box. He cut out her eyeballs and my brother Dee took them outside, where he gave other children, including a precocious nine-year-old bacon-eating "vegetarian," an impromptu biology lesson as they dissected them. Of course the Farm Children didn't take a second look.

When I took to calling her Cyrusina, the lazy way out, it was the nine-year-old who said the name wasn't pretty enough. (I have to agree that it won't be making the top ten baby names any time soon.) Thereafter, our headless pig was known as Lily. In the end, there wasn't a one who had seen Lily boxed and wrapped, naked and pink, prepped and handled, roasted and sliced, her head simmered in a pot, who felt worse off for having "known" her prior to eating her. As for me, getting to know Lily was easier than hauling around the body of Cyrus I, the first whole pig I had roasted in a Caja China. Hoisting his body into a garbage bag full of marinade had given me the creeps. "It feels like Remington," I'd confessed to Husband Hendel, making a connection between this pig and our bloodhound of similar size.

On the other hand, following a field trip to meet some lambs we were to buy, I asked Atticus if he thought it was better to "know" your meat before eating it. "Yes," he replied, without a trace of guile. "It's like they are our friends, only dead." The children proceeded to name every animal as it came to stay in our freezer. Shorthorn the goat, Puffy Legs the lamb, and, in keeping with the anatomical refer- ences, Handsome Trotters, our most recent pig. All it takes to get my children to start eating their dinner with gusto is to drop one of those names, and remind them that this is an animal they know. At first I concealed this practice from Cousin Kate, sure that naming our meat was an urban practice worthy of ridicule. But don't you know she emailed just the other day to say Gimpy, their calf, was destined for the freezer. I guess the main difference is they get to name their meat while it's still alive.

They're like our friends... only dead!

2

. . . .

WHY BUY ANIMALS
BY THE HALF INSTEAD
OF BY THE POUND?

I know, I know, I already said nepotism alone wasn't going to get me a pig. This one I'd have to figure out on my own. That's when Cousin Kate called. Turned out a fellow farmer from down the road had been by, talking weather and whatnot, and he brought up that he had some hogs that were ready to go. Since Cousin Kate's freezer was empty, she said she was interested, and when she told me, well, golly, I was too.

Half a hog sounded very doable, for both of us: a "starter pig" for me, and enough pork to tide her over until the fall, when she planned to share a 4-H hog with one of Jason's colleagues (see the material on 4-H in Chapter 4, "How, Exactly, Do You Buy a Whole Animal?"). Splendid. The farmer raised cows and also grew barley to feed his pigs. I was in.

But here's the rub: While Cousin Kate did all the hard work of arranging the deal, her country brethren don't grow pigs close to *my* home; they grow them close to hers. In other words, this pig was about two hours away. If you are an urbanite hankering for a whole animal, whether you live in Chicago or Portland the chances are good your animal's not being raised just down the road. There's a reason that your co-op charges so much per pound for that shrink-wrapped meat. They've got overhead, sure, but also that pig had to travel. And while ordinary pigs travel distances *far* longer than the

117 miles that mine would take to my freezer, the trip is not something to be discounted. Ingrid, the business partner I sweet-talked into making my virgin pig pilgrimage with me, had barely gotten on the highway when she looked at the address on the GPS and said, "Couldn't you find a pig closer to Seattle?"

Well yes, I could, and since then I have. But for now I'll just note that if part of your reasoning for buying whole animals is to care for the environment, then taking a day off work and chugging along a freeway in a car big enough to hold one might give you pause. Even Cousin Kate said she wondered about the emissions put out by all those sad-looking, smoking vans carting organic wares to farmers markets on a weekly basis. The irony was not lost on me, but I crunched the numbers, and I'm relieved to tell you that even driving a "family vehicle" two hours each way to pick up a pig grown by a small farmer puts you on the good side of Al Gore. (As Jason pointed out, those saved trips to the grocery don't hurt either.)

According to a peer-reviewed study that appeared in the USDA's *Agricultural Research* magazine, the simple act of skipping commercial animal production more than offsets the emissions put out by your car as you drive to fetch your beast. Looking just at cows, and factoring in CO_2 emissions, methane and nitrous levels, ammonia production (in manure), the leaching of important nutrients from the soil, erosion, and the effects of field runoff, the study found that pasture-raised cows contributed 8 percent less greenhouse gas and 30 percent less ammonia than factory-farmed animals. And this doesn't include the processing, shipping, packaging, and selling of said meat. If it still just doesn't seem right to you, find the nearest producer/processor and arrange to pick up your meat at a drop site. Or ride your bike to work for a year to offset those emissions (and work off all that bacon).

You will note that I cajoled a friend into making the trip with me. (Husband Hendel had declined to go, stating that spending that much time getting meat would make him feel "resentful.") If you go to pick up a whole or half-beef or -hog, you'll want someone to help you. Not only does it make the trip that much more pleasant, but those things are heavy.

If you didn't grow up on a farm or know people who buy or harvest meat by the whole animal, consuming meat this way may seem strange—too esoteric, too hard, too old-fashioned, and—in general—too much pressure. But if you've ever brought your own bag to the store or frozen extra food to preserve it, you may appreciate my spiel.

I think many of us can agree that most meat produced in this country is done so in a way that is kind of gross. I knew that back in 1988 when I simply stopped eating meat one day, and if anything it is even more true now with the consolidation of meat production. You don't want to see the sausage being made, as the saying goes, and if you delve into the subject you'll know why. Poorly produced meat isn't good for animals, it isn't good for our bodies, it isn't good for our kids, and it isn't good for our ecosystem.

The mere act of raising food takes a lot of energy and creates a pretty big carbon footprint. The food system accounts for over 10 percent of overall energy use in the United States. Globally, some studies peg livestock production as accounting for nearly one-fifth of the world's greenhouse gases, more than all cars, trains, and planes combined. Reducing or eliminating meat consumption has long been touted as a way to reduce our impact on the land and help feed more people, but let's face it: The world isn't going vegan tomorrow. And if you insist on putting some animal on your plate (as I now do), supporting good meat production is a step in the right direction, providing economic stability for farmers, giving us cleaner air and water as well as less disease and more humane meat, and benefiting every "body" involved.

Buying whole animals lets you choose those that were raised in ways you feel good about. You can get past the buzzwords and the marketing flim-flam and talk to actual people who raise actual animals that will become your actual dinner, for months to come. You can vote with your pocketbook and support those producers raising meat responsibly, however you define it. With a little research, I guarantee you'll find exactly the meat you want: heritage breed meat, meat fed only grass, meat that ate organic barley and cranberries, meat killed by a certain method, meat raised by a future farmer....

Are you getting my point? This matching up of growing practices and values is good for you, the consumer, and good as well for Uncle Dave's cow and all the other animals that end up in our freezers. Yes, here's the spoiler: the animal always gets dead in this scenario, but living a nice life until being dispatched with empathy (and at the very least speed and skill) is certainly better than the alternative.

Is This a Book about Sustainable Meat?

I bought Michael Pollan's celebrated book *The Omnivore's Dilemma*, in hardcover even, and read it while vacationing with my family at the seashore in Oregon, during an expansive moment of space and time that allowed me to really think. I found some of it so fascinating, including the pages devoted to corn sex, that I read portions of it aloud to my teenage nephews. It was such a wonder, a tiny miracle that corn reproduced at all! And to think, it went on to take over the world.

As a former vegetarian I felt I was revisiting familiar ground. I didn't like the way a lot of meat was raised when I was young and sassy. Had it gotten any better? Worse? Was I being duped by clever "greenwashing"? Was I willfully duping myself? Even given my stalwart vegetarian past, I found some of Pollan's book shocking. It was well written; it tugged at me as a reader. I experienced many of the emotions I think I was supposed to: horror, disgust, empathy, fear.

But more than anything, I found the book deflating. With each chapter I would turn to Husband Hendel and declare the further limitations we would need to impose on our diet. "Oh, no more grass-fed beef." Chapter. "Now we can't eat free-range chickens." Read a chapter more. Sigh. "No more eggs." Chapter. "Shopping at Whole Foods is out entirely." And so it went, until we weren't eating responsibly unless we were growing and hunting our own food in a manner disturbingly reminiscent of *Lord of the Flies*.

Pollan's book spurred a whole genre of foodpocalypse writing, including novelist Barbara Kingsolver's book *Animal, Vegetable, Miracle*. This nonfiction sermon quaintly chronicles her family's exodus from the air-conditioned badlands of Arizona to a bucolic Virginia farmstead, where Kingsolver's lush royalty income and a solid ten-month growing season allow the family to test the brutal boundaries of local eating. *Wah*. This book evoked in me not so much depression as mild rage. *What?* I know, I know: I'm a liberal Seattleite who works in book publishing. I realize I have just committed high treason (in addition to killing any possibility for those long-shot blurbs). Hating Barbara Kingsolver? That's like drowning kittens, or buying Chilean blueberries in December.

Please read carefully. I didn't say I *hated* Barbara Kingsolver, just as I don't hate Michael Pollan. Each has a message to get across that, generally speaking, I am in agreement with. The reason for my ire is that my income will not come close to funding a year spent growing or hunting my own food. They have a name for that job, and it is a job, a more-than-full-time job at that. But I am not a farmer. I live in a pretty big city and already have a full-time career, and a commute, my own business, a husband who works sixty hours a week, two young sons, and three dogs. I volunteer in my community, grow as much backyard produce as my Pacific Northwest climate permits, and *still* feel pressured to be a self-sustaining organic farmer in my spare time. The closest I've come was when I lived alone in graduate school and could only afford to eat lentils. Though I understand their point, these polemics, and the discourse in general, have had the opposite of the authors' intended effect on me, inducing nothing but a vague hopelessness. What has happened to specialization? Are we all also to be dentists, and cobblers, and our own surgeons, too?

I am not alone. There are a lot of people out there like me, especially city dwellers, with all our shortcomings and contradictions. I love cooking and raising food, but there are limits to what I can or am willing to do. I have willingly traded keeping food consumers (dogs) for keeping food producers (chickens), for example, because I like to snuggle my hounds, and they would eat the chickens before they could lay a single egg (the neighbors have—had—urban chickens who liked to visit). I will cut up a carcass if need be, but I don't want to kill and skin my own animals on a regular basis, the former being a big downer and the latter requiring skills I don't possess. I will absolutely commit to composting, even if that compost may only produce three $49 tomatoes. The reality is that most days I am lucky to be wearing clean clothes, not to mention making sure my children eat a good breakfast.

So here we are, living a busy urban life. It doesn't excuse any of us from being thoughtful participants in the world and understanding the impact of our choices. I think it's a problem that we trade knowledge for marketing terms and go on about our smug ways. Even if I began this journey investigating ways to buy "sustainable" meat, I ended up buying meat I could believe in, in a way that my family can and will sustain, and that small step feels huge. I found one thing I could understand and do and could tell others about doing, and that seemed to make a nice difference.

There is nothing new about buying animals one at a time and eating them over many months. What is new are the considerations we now bring to the purchase, the informed choices we may make, and the new link that city consumers and rural producers have forged, in no small part through technology.

A by-product of this discovery was a realization that, like most complex issues, growing and buying meat involve a lot of trade-offs. People buy whole animals for all sorts of reasons. For some, buying a half-beef or a whole lamb is the only cost-conscious way they can eat meat produced the way they like, or simply put meat on the table at all with any frequency. For still others, it's a way to ensure the meat they serve their kids doesn't contain chemicals or hormones.

People also *raise* animals for all sorts of reasons: because they love them, because they want to work outdoors, because their father did, because it's what they know. I'll tell you, it's not for the paid vacations. A key theme of this book is accessibility for all types of people—Idea People, Detail People, Urban People, Suburban People, Animal Eaters, and Animal Growers.

Buyers of this book might think: *I* have nothing in common with, let's call them, the Conventional Farmers Who Hate Michael Pollan. After all, we know what meat we want to buy, right? It's grass fed, and organic, and fair trade, and humanely raised, and local, and small scale, and artisanal, and inexpensive, and available year-round, and proceeds from the purchase go to help women entrepreneurs in Tibet, right? When this all began those were the sort of notions I had. But then I started talking to the different people who raise these animals for us to buy and eat.

Here's an example: Unless you're from Washington you have probably never heard of The Evergreen State College. It's a state school of varying reputation, though in my day it was generally known for housing and (somewhat) educating "hippies," as my brother Jeff would call them. There are no grades at Evergreen, and many "Greeners," as the students are called, earn credits for things like tapping sugar maples and skateboarding. I should mention that Husband Hendel has a degree from Evergreen, in mathematics, and he reads physics and engineering books at night before bed; he's that kind of person. So there's that. But he still remembers working through dense topology books, his brain turning somersaults, while his friend spent the year engaged in photography and earned just as many credits as he for her photos—all six of them.

So it was perhaps this bias that I carried with me to a meeting with Evergreen students who raise lambs to sell. I didn't think they would be flakes, but I expected them to be standard-bearers for all the things liberal urban people think we want "good" meat to be. Of all the methods I had discovered for procuring whole animals, from bidding on them at the fair to ordering them online, this was the first college club I had encountered that sold meat, and I was interested enough to want to know more. Since Evergreen is not so far away from Cousin Kate's home, I took Romeo and Atticus with me and promised them a visit with the Farm Children after I conducted my interview. (They do so love the Farm Children!)

As the boys set off to play in the pasture, I met the gathered students and their advisors. We huddled in the soft rain to talk sheep, and they asked me what the book was about. I think I answered that it was about buying sustainable meat, one animal at a time.

But instead of instant liberal approval, I got some slightly defensive fast talking about decisions they made as they raised their animals, and clarifications as to what buzzwords would or would not qualify to describe their lambs.

For example, their lambs were not organic. As a group they had wrestled with the issue and come to the decision that if an animal needed treatment for an infection, they would give it antibiotics. Even a single use of antibiotics takes an animal out of the organic category. While they fed the lambs only grass, they had found that their ewes needed extra nutrition that grass could not provide, especially when pregnant and just after lambing while they were lactating. First they tried giving them dried cranberries, but they finally decided to supplement the feed with grain. As someone who has been pregnant, I assured them that my sympathies rested completely with the ewes. If they needed more food, for Pete's sake, give it to them! They used locally grown oats and barley in order to support the local economy and reduce the energy needed for transport.

It was in talking with this group, with Jesse and Alex and Alea and Mike, in discussing the fortitude and knowledge that it takes to raise animals and to do so responsibly, that I settled on a better benchmark for what was good and right as I searched out whole animals to eat. They critiqued the elusive definition of "sustainable" and instead strove for what was "reasonable" in the care of their sheep. The real threat to sustainability, as they saw it, was the day that not enough young people would have the knowledge and skills to practice animal agriculture.

Along the way, they were investigating and truth-testing all the accepted conventional ways to grows lambs, from castration to the docking of tails. Not seeing what conventional farmers were doing and then doing the opposite, but factoring in the *reasons* why accepted practices existed and seeing if they could be improved upon.

Case in point: that tail docking. Lambs are born with tails, by the way; they're not like Manx cats. But conventional practice has their tails shortened or "docked," usually shortly after birth. Lambs' tails are not docked for fashion but to prevent fly strike, a disease in which poopy rear ends provide a hospitable environment for wool maggots that burrow into the flesh of the sheep. Fly strike can result in both pain and death for the animal. The sheep club was experimenting with not docking tails, sparing the animals a procedure, but they had also lost an animal to fly strike, and it was not a loss they were happy about. They didn't yet have a definitive answer, but they said the conventional sheep farmers had told them to keep them apprised. The entire process seemed thoughtful, and I appreciated their contact and community with non-Greeners.

Favorable Environment

○ ○ ○ ○ ○ ○ ○ ○ ○ ○ ○ ○ ○ ○ ○ ○ ○ ○ ○ ○ ○ ○ ○ ○ ○

"Wise selection and breeding and the establishment of definite types of animals suitable to special purposes cannot accomplish the desired ends unaided. The feeding and care must receive as much attention as the breeding. No matter how well bred an animal may be, and no matter how great may be its tendency to conform to a given type, it must enjoy a favorable environment before its inherited good qualities can fully assert themselves and thereby enable the animal to fulfill its mission."

—from *Types and Market Classes of Live Stock,* 1915

Finding the Middle Ground

During this same trip Romeo, Atticus, and I arrived at the Chehalis Livestock Market just in time to meet up with Cousin Kate, Jackson, Margarite, and the Pterodactyl Baby. Knowing my meat preoccupation, Cousin Kate and Jason had nicely thought of me and invited me to come watch the auction to see what it was all about.

Trucks loaded with livestock, mostly cattle, were idling in the parking lot, having unloaded their charges. The auction itself hadn't started yet, so Cousin Kate did her greeting of the locals as we passed through to the catwalk, a perch from which we could safely observe the action.

"Look," Atticus said, "cows!" Indeed, city child, cows. Brown Jerseys that looked like teddy bears, some Holsteins with their hips jutting out, and some wild Herefords that had no idea what in the hell had happened: one minute they were out grazing some remote pasture, and the next minute they were in a chute, with a mysterious man (Jason) sticking his (gloved) arm deep inside their business ends—just like James Herriott. Jason works at the market every Friday, examining, vaccinating, and dehorning the cattle that come to auction.

Examining cows like that is exactly as easy as it sounds. In order to keep the cows from moving while Jason examined them, it took three men, a

LOOK! COWS!

chute, stanchions, and some septum pinchers that hooked to a wall. I didn't like the process of getting the animals into the stanchion, and I didn't like the ring. It made me squirm, it looked so uncomfortable. Cousin Kate laughed softly at my furrowed brow and pointed out that the 1,200 pounds of wild steer pinning her husband against the metal bars made him a little uncomfortable as well. Point well taken.

Jason took me and introduced me around to the farmers and the owners. My favorite comment came from Sam, a buyer for a slaughterhouse north of Seattle. He goes to the auction every Friday. He has a practiced eye and knows just which cattle to bid on and how to get the price he wants. He was kind, very gentlemanly, a man who grew up on a cattle ranch in the Midwest and had cows in his blood. He asked me if I'd come to the auction to find out where my hamburger came from. I said it was something like that. He laughed and, with a twinkle in his eyes, said, "Well, it ain't pretty."

Another young farmer came along who had found great success making yogurt. He sold out at the farmers market every week. He and the others were what city people would call "conventional" farmers. They talked about how difficult it was to get organic certification, and what that really meant. They talked about the organic farmers who, when they ran out of hay or feed, would buy some from them to get them through, a move that would technically disqualify their products—whether that be meat or milk—from being termed "organic" according to the USDA. Each camp seemed to be borrowing a bit from the other, everyone trying their best to get by.

Before Joe, the co-owner and auctioneer, got going, another co-owner's daughter read a poem she had written in honor of Veterans Day. Joe asked all the veterans in the room to stand up so we could applaud them; there were quite a few, old men who doffed their hats at the large flag hanging on the wall. And then the auction began. One at a time the cows made an appearance in the ring, accompanied by Joe's mesmerizing rhythmic singsong. He had mad skills; I was not surprised he was the 1987 World Reserve Grand Champion Auctioneer.

I don't know what I expected, but I didn't see anything amiss at the auction. There were some wild cattle, some skinny cattle, and some awfully confused cattle. Cousin Kate said that sometimes the lot overall looked better than others, and some days she and Jason left the market with an animal. That day there was nothing that caught her eye. I found the company hospitable and I liked Sam, but I'd rather eat a burger ground from one animal than many.

This middle ground is where I ended up spending most of my time in considering my purchases. I'm all for more precise definitions of terms that describe our food, but in setting the terms around meat I feel good about eating, I found myself feeling closer to Supreme Court Justice Potter Stewart who, in 1964, so famously said that while he could not precisely define hard-core pornography, he knew it when he saw it.

Later that same night, after feeding the sheep and visiting the auction, talking to the students and chatting with farmers, we followed dinner with Cousin Kate's family with a venture into another charged topic, maybe the most charged of all: politics. Let's just say that in talking over the upcoming presidential race with some of the most iconic figures in this book's Cast of Characters, it became quite apparent that our views break across stereotypical urban–rural lines. And like making hamburger, it wasn't pretty. Once we got past the quips and the zingers and the sound bites, and though we did not get to a place of total agreement (perhaps even "mild agreement" is a rosy revisioning of that evening), we did find certain values and goals on which we could agree—that precious middle ground. And for change to happen, whether in the White House or the barn, perhaps that is the place where we should begin.

MYTH #2

If I buy a whole animal's worth of meat, I will have to eat a whole bunch of *weird* stuff!

Cousin Kate graciously allowed me to give the cutting instructions on the first half-pig that I ordered with her. I would never tell her this, but I had never given "cutting instructions" before, and I was nervous. But by the time I called them I had templates I'd downloaded from the Internet, and I was totally prepared, I thought, for any complicated questions they would ask about parts I didn't know the name of, so as not to reveal that I was nothing but a silly city slicker.

In reality, however, this is how that conversation went:

Processor: How thick do you want your chops?

Me: An inch and a half?

Processor: How many to a package?

Me: Four. (*Easy one!*)

Processor: Do you want mild, medium, or heavy spice on the sausage?

Me: Medium. (*Since she said "heavy" I assumed this was how aggressively it was seasoned in general. In reality, this meant how* spicy *do you want your sausage. Medium, as it turned out, was too hot for my children.*)

Processor: How thick do you want your bacon cut?

Me: One-eighth inch?

Processor: Do you want your hams cut in half?

Me: Yes. And fresh, please, not smoked. (*I was thrilled to know and use the term "fresh ham."*)

Processor: OK, that will do it.

Me: Wait... (*This had been the opposite of the conversation I expected to have.*) Can I have all the fat?

Processor: Um, sure.

Me: Can I have the hocks smoked?

Processor: Sure.

Me: Can I have the cheeks? (*Since I wasn't asking for the ears, I felt remiss in not asking for the cheeks. It seemed wasteful.*)

Processor: Oh, I think the head's already been cut off of your pig. We have some other hogs coming in tomorrow. I could probably give you someone else's cheeks. I mean, if you don't mind.

Me: OK.

Now, this processor didn't give me as many choices as some that I've dealt with since then. The butcher of the pig I ordered after this one let me determine whether or not I wanted steaks or roasts from the shoulder, for example, and didn't blanch when I asked for the fat and the belly whole. However, when I asked if the butcher could scald and scrape my pig instead of skinning it (so I could cure the cheeks and have skin on my fresh hams), it took five email exchanges for them to understand my question enough to say no.

My point here is that *I* was the one who tried to get fancy with my animal ordering. Cousin Kate was surprised when she ended up with fresh rather than cured hams, really just big roasts, but she was game to try them. She told me to keep all the fat—she wouldn't be rendering her own lard, and did not want the cheeks either, thank you. I think if I had brought up the ears she would have disowned me as kin. The "weirdest" thing I ended up with in my freezer in this scenario was a pork steak, which is simply a steak cut from the shoulder. For heaven's sake, that's hardly a thymus gland (which you may know as "sweetbreads").

When it comes to a cow, you have to ask for the tongue or liver if you want them, as many processors don't routinely offer them. Some even charge extra. You can't get tripe from pigs even if you try. Lambs and goats, on the other hand, come fully equipped. You may not check any box that says "offal," and yet each may come with tiny cryovacked heart, kidneys, fat, and liver. The key word here is *tiny*—most of that meat is consummately recognizable and doesn't get any weirder than goatburger.

SILK PURSE

3

. . . .

IS THE WHOLE-ANIMAL
LIFESTYLE FOR YOU?

Whether you care most about the quality and flavor or the conditions in which your meat was raised, or you are trying to support local farmers, buying a whole animal at a time is a surprisingly good choice for many people.

For example, have you dubbed yourself that darling of new lifestyle buzzwords, a "flexitarian"? Now, I would call a person who tries to eat a mostly grain- and vegetable-based diet without completely eschewing meat a cheating vegetarian, but I'm not here to judge. Though this may seem counterintuitive, I have found I eat less meat when I only eat meat that I know, making the whole-animal lifestyle a good fit for flexitarians as well as avowed carnivores. And it's meat you can feel good about to boot.

Do your kids eat nothing but hot dogs? Buying whole animals and cooking them is a spectacular gift to your children, in understanding where and how your food was raised, in knowing exactly what you're getting (and not), and in teaching them about a variety of animal parts and how delicious they can be. Or maybe your kids never eat hot dogs because you're concerned about *E. coli*, pesticides, hormones, or antibiotics, or want to avoid processed foods or artificial ingredients. Well then, how about if you *make* your own? Even if you don't want to get fancy, I will tell you that children love both pounding and grinding meat. Buying whole animals also gives you the opportunity to show kids exactly what goes into sausages, as well as all the parts on an animal we can eat. Okay, when it got to the point that I was separating out the primals on our dog to show Atticus the component meat cuts of an animal ("Here's his *rack*, and here, the *loin*..."), I knew I had gone too far. But I think there's a sweet spot in there somewhere in which your kids don't think meat grows wrapped in cellophane.

For all those Idea People like me, who act before they think and need help making it all the way to the finish line once they've got a carcass in the freezer, I'm here to help. Help you, for instance, to know *before* you buy a whole animal that an average cow comes with a hundred pounds of ground beef and only about twelve New York strips, that a pig's hind leg is called a fresh ham, and that a whole goat will fit in an apartment freezer. This book will give you the answers you need for that important conversation with the butcher that involves the term "cut sheet."

That said, buying a whole animal is not for everyone, romance aside, and not the only way to save the world. Here are some questions to ask yourself, honestly, before taking the plunge.

Are You Equipped to Lay Out Some Real Cash Up Front?

Buying a whole animal costs anywhere from hundreds to thousands of dollars. Though the overall price is usually quite reasonable—the same as, if not lower than, what you'd pay at the market—the fact is that you have to pay it out in one lump sum or put down a sizable deposit. If this is not financially feasible for you, consider going in on animals with friends. Many producers also offer meat packages or bundles, giving you larger amounts than you'd usually bring home from the supermarket, but at a lower price than you'd pay if you bought all the cuts separately.

Are You Prepared to Store It?

Few people living in urban environments are prepared to store an entire beef. Multiple freezers are required, after all, and an emergency generator to protect your beef if your power should fail...not to mention the hound dog you should get if you already have multiple freezers and a generator, just to complete the "look." Some people are cursed with apartment freezers so small you couldn't squeeze in a bag of peas. I know, because I've had that freezer.

Even if you purchase part of an animal or a small beast that will just fit in your freezer compartment, be warned that's probably most of what your freezer will contain, and barely at that. No one wants a frozen roast landing on their toes every time they get some ice, and I don't want you to grow to resent rather than appreciate your animal once it comes to live with you. If you're space challenged you could go in on an animal with a friend and store it at their house, but that leads me to my next question.

Will You Really Cook It?

The Natural Resources Defense Council found that each of us throws away about our body weight in food waste each year. In general, I would offer that buying meat by the animal gives us a chance to decrease that waste considerably. Not only is there an opportunity to use more of the animal at the time of slaughter, but because your work sourcing and buying the animal is so intentional, there will probably be far less forgotten meat languishing in the back of the fridge. No more peering through the plastic, poking with your finger, wondering if the meat you bought last week at the market is still good. For me, all the meat I buy is special—the children named it, after all. I cook it, and eat it, and cherish it. Throwing away even leftover leftovers breaks my heart.

That said, frozen meat is not instant food. It takes time, sometimes days, for meat to defrost in the refrigerator. Don't buy a whole animal if your schedule will have you eating takeout every night next to your very full freezer. Likewise, if your family won't eat anything but steak and hamburgers, or lamb chops—*frenched, please!*—and you're not willing to push their boundaries, then the whole-animal lifestyle is not for you. Instead, try to buy those traditional cuts you crave from a local producer that raises animals the way you like.

If any of the above restrictions has you rethinking your ability to handle a whole animal, there is still nothing stopping you from trying any of the recipes with good meat procured a different way. You might also find that, having the information, you feel compelled to educate others on the whole-animal process. In the course of my research I found a grower I really respected, but I already had a lamb in the freezer. I went so far as to become a meat pimp, and handled the instructions and money and even fetched the thing, all to then pass it on to family and friends with nary a chop staying put in my home.

People Are Even Nicer Than Pigs

○ ○ ○ ○ ○ ○ ○ ○ ○ ○ ○ ○ ○ ○ ○ ○ ○ ○ ○ ○ ○ ○

Do you want to know a really good reason to buy a whole animal? People. People who grow food and animals are, by and large, nice. People in general, actually, are nice. Watching political debates and reading the newspaper, even going to community meetings sometimes, one can lose sight of that fact.

I know I can. I had already had a hectic morning, and now I had to drive three freeways to pick up half a pig. I was grumpy. I couldn't even get excited about the meat, which usually would have cheered me up, because while I would soon have sixty pounds of delicious pork, I lacked a working kitchen in which to cook it. Tromping through drywall dust to fill my freezer with meat I wouldn't get to taste for who-knows-how-long didn't feed my immediate-gratification-seeking spirit.

It got late, and I was lost. I was supposed to meet Martin within a half-hour window outside the city in an Embassy Suites parking lot. I drove around in a sea of strip development and warehouses, looking in vain for his van. If this sounds like a drug deal gone bad, well, it had started to feel like one.

And then I spotted him. I pulled over and opened up the back of my SUV. He hopped out of his truck smiling. And you know what, he was lovely. We chatted as I packed pig parts into my ice chest—about my meat research, about my brother who had picked up the other half just two days before, about their farms' pigs and chickens and cows, and the meat birds they planned to add to the mix. He told me how they were expanding the operation to add thousands of chickens, but all in a huge cold frame so the birds could run around and eat grass year-round. He gave me a complimentary dozen of his eggs to try.

We talked about the meat business and about farmers markets in Seattle that might be possibilities for him. Knowing I was curious as to how his Berkshire breed would taste in comparison to plain ol' pig, he said these pigs should be especially delicious as they had been fed the apples from Johnson Orchard in Yakima. "Johnson's!" I exclaimed, excited because the orchard is just minutes from my childhood home. Every year in late summer I still go and buy boxes of pears and apples and peaches while I'm visiting my family. I felt less than six degrees now separated me from Handsome Trotters, the name Romeo would bestow upon this, our latest pig.

I drove away smiling like I'd just scored some E, already planning how I would cook this apple-fed piggy and buoyed by Martin's smiley face and infectious positivity. Martin used to be an architect, but he loved working outside. More than that, he loved getting out there and talking to people about his products and how they were produced. He really did. Not only did I feel great about eating the meat, I was glad my dollars had gone directly to support this small farm.

People. They're even better than pigs.

MYTH #3

I don't have room in my freezer for a whole animal!

In truth, you might not. A cow is big; so is a pig, actually. Or an elk. If you're trying to stuff a whole animal into the freezer compartment of your fridge, you're going to have to plan. But there are ways around this:

1. Don't buy the whole thing. Easy, right? If you are outfitted with a fairly standard refrigerator/freezer arrangement, I can get you down to an amount of meat that should fit: an eighth of a beef or a half-pig. Buy less from the producer, or practice "cowpooling" as many charmingly call it, sharing a purchase with friends. Or buy a smaller animal. So you grew up eating beef. You know what? Goat is delicious. So is lamb. And they are little animals. You could practically store a whole goat in your underwear drawer (if it were really cold, I mean).

2. Clean out your freezer. In preparation for moving our current refrigerator/freezer from the soon-to-be-remodeled kitchen, I cleaned the darn thing out. Since we have a separate meat freezer in the basement I didn't really bother so much with this freezer on a regular basis. As a result, do you know what I found in there?

    Two baggies full of half-grated nutmegs
    Cheese rinds for stock (over a year old, if not two)
    Chicken parts for stock (over a year old, if not three)
    Fourteen bags of edamame
    A Costco allotment of Otter Pops
    Many, many Chinese noodles for soup
    Frozen banana pulp (two containers) for bread I never made
    Frozen banana leaves for steaming fish
    Mochi
    Gluten-free waffles from when Romeo was thought to be allergic

Freezer-burned "natural" juice pops (when the kids prefer
   Otter Pops anyway)
An unidentifiable whole fish
Three bags of frozen peas that may or may not have been used as ice packs

I won't go on, but you get my drift. Use your freezer the way your freezer was intended to be used, instead of as a very cold junk drawer. You may find that you waste less food and you have more room than you thought. I know I did.

3. Pay a storage "fee" to a friend with a freezer. Your fee might be a package of chops or bacon, or an invitation to lamb stew night. Meat can build community.

butcher's parcel
" old timey and picturesque "

4

· · · · · · · · · · · · · · · · · · · ·

HOW, EXACTLY, DO YOU BUY A WHOLE ANIMAL?

In the pages that follow, I explore the many methods, ways, and people involved in buying meat by the whole animal, and offer animal-specific instructions in each section. However, before you begin you might want to know, in broad terms, how this whole thing works. When my involvement began, I didn't. Here's a primer.

Step one: Determine your meat habits and values. First, assess what kind of meat you usually eat and know how to cook. Consider the general meat consumption of your household. Buying a whole or part of an animal may change these patterns in the end—you might get a hankering for goat, for example, or find that you love ground lamb when you had never cooked with it before, or discover that eating habits change when you commit to eating only the animals you have bought—but we all need to start somewhere. Think about whether you want an animal fed a specific diet, or a particular breed. This is also the time to talk to family and friends and see if anyone wants to go in on an animal with you. Based on all these factors, decide what kind of meat you want to buy.

Step two: Find a grower or a producer. The grower or producer is the farmer, the one who raises the animals—the animal agriculturist, if you will. I use these terms interchangeably throughout the book. In the pages to follow I'll tell you how to find growers in your area for the animal of your choosing.

Step three: Communicate with the grower or producer. That's why you're buying meat this way—so you can talk to them! Contact them, and find out what their animals eat, what breed they are, how big they get, how they are butchered, and how much they cost. In each animal section I give tips on questions you might want to ask.

phone (for communicating with growers)

Step four: Reserve your animal! Especially from smaller growers, supplies are limited. Whether you want a 200-pound hog all to yourself or need the grower to match you up with a partner to buy half a lamb, you'll want to look into your meat well before you want it. You pay a deposit to reserve your animal.

Step five: Figure out what cuts you want. Run through all the types of cuts you can get from your animal and products that can be made from the meat, including cured and preserved meats. Your wishes as to how your animal should be cut and which parts you want to receive may be made by phone, email, online form, or through a document called a **cut sheet**.

Step six: Communicate with the processor. The processor is who city people think of as the butcher, and that's not wrong. He or she may or may not be the same person who kills your animal, but this is the person or facility that will cut up the carcass, then vacuum-seal it or wrap it in paper. This is called the **cut-and-wrap**. The processor may also prepare some of your meat, say, by having your bacon smoked for you or by making it into sausage. You need to determine and discuss what is on your cut sheet so that you end up with the cuts of meat you want in the amounts you'll use. You may communicate with the processor directly, or through the producer; your producer will tell you which.

Step seven: Pay up! You may pay only the producer, or you may pay the producer for the animal and the processor for the services provided. Again, your producer will tell you what to do.

Step eight: Pick up your meat. You may pick it up at the processing facility, from the producer, or at a prearranged location, depending on the circumstances. We'll further explore these options later.

How Do I Find My Own Animal?

You don't have an Uncle Dave, you say? Well, fortunately for you there are many, many options for sourcing and buying your own whole animal. When I embarked on this journey I had no idea of how one might go about this if your kinfolk didn't raise food, but by the time I got going I realized it was easier to buy a whole pig than trim my own bangs (with much better results). The options run the gamut from highly involved to effortless, depending on where you live, what type of animal you're looking for, and whether you want part of an animal or a whole one. Price varies quite a bit, though in the sections devoted to specific animals I give ranges for approximate costs. Warning: Cousin Kate tells me beef prices are on the rise—good for Dave, less so for consumers. Please remember that costs fluctuate.

Farmers Markets

For all the reasons you love buying your fresh fruits and veggies at your local farmers market, this is a nice option for sourcing a whole or partial animal. Every grower I spoke to who was selling cuts of meat at the market also took orders for whole animals, requiring little more than a week or two of lead time to deliver a whole custom-butchered goat or lamb in season (pigs and beef have specific slaughter dates). Even if they don't advertise whole animals, it doesn't hurt to inquire. I ordered my first goat at a stall that primarily sold goat's milk for drinking and cheesemaking. "Whole goats for sale" wasn't written on a sign, but all I had to do was ask.

The farmers market is an easy option for buyers for whom organic or grass-fed meat is important. Market growers who come to hawk their meat already have a sustainability angle going, because it's important to the demographic. You should always ask for the specifics, but I'll bet you that "grass-fed" or "pastured" is already up there in 60-point type on their banner. White folks with dreadlocks and purveyors of handmade herbal soap won't settle for less.

If you're feeling nervous about ordering a whole animal, the farmers market option allows you to meet your grower in person and talk to him or her. Generally I find it easier on the nerves to make a fool of myself in person than over the phone. Farmers like to talk about what they do, whether they raise rutabagas or Boer goats, and if they have the time you can find out everything you want to know about where your future taco lived out its childhood. But remember, the vendors are there to sell their products, and they have customers other than you. If you have a laundry list of questions to go through, plan your visit accordingly. I have to say, there's something about that initial contact that is meaningful; part of buying a whole animal is forging a more personal connection with your food and food sources.

Convenience is another major plus for sourcing meat at the farmers market. You'll pay a bit of a premium, but after having trekked nearly 130 miles to pick up my own meat from one processor, driving or walking down to the weekly market to pick up my meat sounds awfully nice. A warning, though, if you walk: You're going to want to employ friends or a Radio Flyer or two to help get your meat home, or ride a well-equipped Xtracycle with panniers. (On the plus side, your urban sustainability rating for this method is seriously amplified.) It also depends on what animal you buy, of course. Romeo and Atticus handily wheeled a whole goat around the farmers market in a single ice chest, while I could barely budge the same chest full to the brim with half a pig. (See the Cow, Pig, Goat, and Lamb chapters for specifics on how much meat to expect.)

Buying animals from these growers enables you to try before you buy, to sample their steaks or chops or roasts ahead of time, to see if you really love the flavor, texture, and marbling before you commit. If you plan on having the processor make sausage for you or smoke your bacon, sampling first might help you avoid the horror of the Tremendously Over-Salty Sausage I received from one processor.

The drawbacks of sourcing your meat at a farmers market are largely rooted in cost. Usually, buying a whole animal saves you money based on economy of scale. Even factoring in grass-fed, organic, or pastured meat, you're certainly still going to see some savings if that's all you buy anyway. But I found that the growers at urban farmers markets by far had the highest prices. For example, here is a breakdown from one fact-finding trip to one of the largest farmers markets in Seattle:

I bought one small boneless leg of grass-fed lamb for $37 at a little less than $9.50 per pound. The same grower sells whole lambs for $450 each (hanging weight about 55 pounds) including slaughter, custom cut-and-wrap, and delivery to any market they go to. For lambs, figure on about 65 percent of the hanging weight in meat, give or take depending on whether you like your meat boneless. This brings us to just over 35 pounds of lamb meat, averaging out to about $12.50 a pound. Now, if I were to buy lamb in the butcher shop, I'd pay about $4.99 a pound for shanks, but $19 a pound for rack of lamb. But I wouldn't know as much about that animal as I do about the first.

It doesn't even have to be your local farmers market. I bought a pig from some lovely growers when I was visiting the farmers market where my family lives, figuring I could make the drive and see them when I went to collect the pork. But by golly if they didn't deliver my pig to a local drop location, all for an extra five bucks.

4-H Animals

Go to the fair! I don't know what kind of childhood you had, but in the bustling metropolis of Yakima we loved our fair. The Central Washington State Fair came to the fairgrounds every September, and it was enough of an event that in grade school we got out at noon for Fair Day, and everyone collected coupons that came with bread from the local bakery, Snyder's, to get in for half price.

I was never a huge fan of the rides (I've always had a delicate inner ear), so what I remember most are the greasy, crisp elephant ears that left cinnamon sugar detritus on your face and fingers and, more than anything, the smell of the gorgeous barns, the sifting autumn light, and the clean smell of hay and the lop-eared bunnies, bearded goats, handsome horses, and doe-eyed calves. Perhaps it was that experience of petting those charming animals, coupled with the sugar high of cotton candy, that produced in me my current longing for eating the cute animals. These were 4-H animals, and though I was never country enough to raise my own, God bless the children who do so now, learning responsibility and a courage in letting go that I don't have. I'm pretty sure if they saw how I treat my dogs (like small children), they'd laugh at me.

If you have a county fair near you, you have access to perhaps the most wholesome way to buy a great animal and support future farmers at the same time. For you true city folks, 4-H is a youth development organization that helps kids develop skills through real-world, hands-on practice. The four H's in question—Head (Managing, Thinking), Heart (Relating, Caring), Hands (Giving, Working), and Health (Being, Living)—are values the kids focus on and develop through different programs, such as buying and caring for an animal and raising it to market weight, then getting the best possible price when selling it. A stellar example of the club at work, Jason credits 4-H experience with leading him to veterinary medicine.

A staunch supporter of 4-H, Jason's veterinary practice buys a 4-H pig every year, and Cousin Kate has started tacking on an additional lamb. I had my hopes pinned on buying a lamb in Yakima at the fair of my childhood, but I found out that state fairs don't work the same way. Check out your local fair for details, but in general this is how it works for a lamb. The lambs are born in February to early March, generally speaking. The kids (human ones, not goats) buy the lambs in April or May and then care for them and feed them all summer until the fair auction in September. To reach market weight, the lamb should be somewhere between 100 and 140 pounds, the same weight a farmer would target in marketing his or her lambs to consumers. The children send out letters or stop by businesses, trying to find customers to buy their animals. Being cute doesn't hurt, for the kid or the animal.

When it comes time, judges look over the 4-H animals and hand out ribbons—white for overly fat animals or ones with inadequate finish, blue for healthy animals that can be auctioned. In the last days of the fair you sign up for a bidder number. Floor price for the auction varies widely, depending on the species and the commercial market prices for the year. Cousin Kate reports that this is the best lamb she buys, so tender you could cut it with a fork.

Google It

Lucky us—we live in the age of the Internet. You can find a whole animal in about forty-three seconds on the Web. Now, it will still involve some back and forth in terms of knowing what you're getting into: hanging weight, live weight, knowing your primals and your cutting instructions, and what you'd like to pay. But there are nice growers out there who have taken the time away from the tractor to put up a website just for people like you. Some are fancier than others, some are little more than a different way to deliver their phone number to you, but search and ye shall find. For an extra fee, many growers and producers will also ship your meat to your door. It's like Amazon, with cows.

internet research
localharvest.org

The downside of this? The lost personal connection. I bought a lamb through a grower I found on Localharvest.org, a fabulous online resource that connects you with local farmers and producers. I went to the grower's website, which told me everything I needed to know, and I sent an email reserving a lamb. He sent an invoice when the lamb was processed, and I went to the butcher to pick it up. I never met the grower, and as I drove through the rolling green hills to the butcher I couldn't help but wonder which pasture Puffy Legs had gamboled about in just a week before. It was very convenient, and the only effort required was picking up the parts at the butcher. But it was also a little clinical and impersonal. I missed the talking. I missed the people. I found I preferred a combination approach, using the Internet for research or email for giving cutting instructions but meeting the grower somewhere along the way.

Craigslist

I would love to tell you whose idea this was: Uncle Dave's! I didn't quite believe it when Uncle Dave said it, but gosh darn it, he was right. Farmers with meat available often advertise here. Try searching under "whole beef" or "hog." Buy a used couch while you're at it. There are similar sites, some of which specialize in foodstuffs. In the Northwest, for example, there is FoodHub (food-hub.org), while eatwild.com connects you with pastured animals around the country. You didn't know it was that easy, did you?

Your Favorite Restaurant

Do you love the steak you get at Chez Locale, that "named" steak from Painted Boat Oar Locks farm? Track them down and see if they'll sell you the whole cow. This is a sneaky way, like checking the acknowledgments in your favorite novel to get the name of a good agent or editor. But it works.

Isn't All Meat "Natural"?
Meat Marketing Terms Demystified

There is a dizzying number of adjectives used in the marketplace to describe your potential meat, from "grass-fed" to "natural" to "pastured." Some of you won't care what your animal ate or where it had its breakfast; buying meat from a local farmer or for the right price might be all you need to go on. However, for you Detail People, below I attempt to demystify these marketing terms one by one, so that you can make informed decisions about the type of meat you and your family will enjoy the most. The Cow, Pig, Goat, and Lamb sections of the book go into further detail about these terms as they apply to each specific animal.

Grass-Fed

Perhaps one of the most flaunted terms in the whole animal world is "grass-fed." Though it seems like a simple concept, this term has elicited some consternation because in the past, "grass-fed" animals were allowed to eat grain in addition to grass. Then, in 2007, the USDA finally weighed in on a more precise definition of "grass-fed ruminants" (this includes cattle, sheep, goats, and giraffes, but just try to get giraffe neck into your freezer!). What is that definition, exactly? I'm so glad you asked. I reprint some of it here not only for clarity and transparency but also because I love the terms "haylage" and "baleage":

> USDA Definition of Grass- (Forage-) Fed
> Grass and forage shall be the feed source consumed for the lifetime of the ruminant animal, with the exception of milk consumed prior to weaning. The diet shall be derived solely from forage consisting of grass (annual and perennial), forbs (e.g., legumes, *Brassica*), browse,

or cereal grain crops in the vegetative (pre-grain) state. Animals cannot be fed grain or grain byproducts and must have continuous access to pasture during the growing season. Hay, haylage, baleage, silage, crop residue without grain, and other roughage sources may also be included as acceptable feed sources. Routine mineral and vitamin supplementation may also be included in the feeding regimen. If incidental supplementation occurs due to inadvertent exposure to non-forage feedstuffs or to ensure the animal's well being at all times during adverse environmental or physical conditions, the producer must fully document (e.g., receipts, ingredients, and tear tags) the supplementation that occurs including the amount, the frequency, and the supplements provided.

Now, the USDA isn't peeking into the pasture of every small grower in the United States, ensuring that his or her animals aren't nibbling a little corn in its post-vegetative state. However, if you buy into a whole animal and buying an exclusively grass-fed animal is important to you, in most cases it's quite easy to go to the farm and verify the animal's diet and environs for yourself. For me, the really great thing about buying into a whole animal has been not feeling the least bit compelled to check up on my farmer; researching a grower and then communicating directly with him or her is enough for me to trust their claims. The ones I've dealt with have all been quite proud of their methods and of how they raise their animals. (Being related to a grower doesn't hurt either.)

Grass-Finished

What in the heck does "finished" mean? Animals are "finished" before they are finished off; "finishing" describes the time when animals are fattened up before slaughter. A "grass-finished" label indicates the animal was not finished on grain, which alters the nutritional profile of the meat.

Pastured

If you thought pastured meat was the same as grass-fed, then you wouldn't make a very good attorney. It's all in the fine print. As you can see in the excerpt from the USDA guidelines, the government has come down on *what* ruminant animals are allowed to eat in order to be deemed grass-fed, but it stopped short of dictating *where* they should eat. For some consumers, even if cattle or sheep have "access" to pasture, feeding livestock hay in a barn or feedlot is not the same as an animal foraging for its own food at home on the range, so to speak. The term used to describe this kind of freewheeling animal lifestyle is "pastured."

I investigated one easy-to-spot verification system that I encountered not infrequently when sourcing meat, especially beef. The American Grassfed Association (AGA) is an independent group that developed standards "intended to incorporate the attributes of open pasture, animal welfare, no antibiotics, no hormones, and the production of nutritious and healthy meats and to be supportive of American family farms." All seems good there. They also specify that grass-fed animals must be raised on pasture and may never be fed grain:

> **From the American Grassfed Association Standards:**
> **3.1 Forage Protocol**
> » All livestock production must be pasture/grass/forage based.
> » Grass and forage shall be the feed source consumed for the lifetime of the ruminant animal, with the exception of milk consumed prior to weaning. The diet shall be derived solely from forage consisting of grass (annual and perennial), forbs (e.g. Legumes, Brassicas), browse, or cereal grain crops in the vegetative (pre-grain) state.
> » Approved supplements may be fed as outlined in section 3.3.
> » Animals cannot be fed grain.
> » Animals must have continuous access to pasture and forage appropriate to the species.
> » Forage is defined as any herbaceous plant material that can be grazed or harvested for feeding, with the exception of grain.
> » All AGA grassfed, pasture finished and dairy animals must be maintained at all times on range, pasture, or in paddocks with at least 75% forage cover or unbroken ground for their entire lives.

If you read through their extensive standards, however (and I did so you wouldn't have to), you will also note that the AGA "further recognizes that the US is geographically and climatically diverse and that grassfed production without any supplementation may not be feasible in some regions of the country. In developing a pasture finished standard the AGA brings together farms who share the attributes stated above but may differ in their approach or need for supplemental feed on pasture."

But wait—didn't we just learn that grass-fed animals only eat grass? Isn't this the American *Grassfed* Association? Confused? You're not the only one. According to AGA standards, a specified lifetime percentage of the feed for AGA certified cattle may also include treats like canola seed and meal, malt sprouts, and soybean or coconut meal and other foods that are nearly all what are called by-product feeds (e.g., canola meal is a by-product of extracting oil from canola seeds). According to my resident expert (Jason), most of these are less digestible than grain but are cheaper and higher in protein and fat, making them useful to some growers during winter months to supplement mature pasture or hay.

So, where does that leave us? If AGA's standards meet your personal standards, and the association logo makes it easy for you to choose the meat you want, just remember that these growers pay for the privilege of that user-friendly marketing tool. They also must meet every criterion laid out in the AGA standards for eight of ten years of participation, or they are dropped. As with many noncertified organic farms, smaller-scale producers may agree with the methods but find the certification too difficult and costly to pursue.

Organic

Even if a "certified organic" stamp normally elicits in you a Pavlovian response, prompting you to toss said product straight into your undyed-hemp reusable grocery tote, you should understand precisely what "certified organic" means in relation to meat. For people who want grass-fed and -finished meat, then just looking for an "organic" label won't do it. Your animal could have been fed grass, grain, supplemental feed, or a combination, as long as the feed was grown according to organic standards:

Regulations set forth by the National Organic Standards Board and USDA provide these certified organic standards:

» prohibit the use of any antibiotics or growth hormones in organic meat and poultry,
» require 100 percent organic feed for livestock,
» prohibit the use of pesticides and/or chemical fertilizers in the growing of feed,
» stipulate that the land on which grass or feed is grown has not been sprayed for a minimum of three years,
» prohibit the use of Genetically Modified Organisms (GMOs) in livestock feed,
» do not permit animal by-products in livestock feed,
» require third-party verification and inspection,
» require a special processing plan for processing and handling the meat.

As the example of the Evergreen lamb club made clear, animals considered organic cannot receive antibiotics even to treat disease. Parasites can be a problem for organically raised animals as well. Some producers complain that staying in compliance with organic certification means putting a lot of time and effort into planning, documenting feed sources and systems for harvest and storage, testing soil and water, submitting to annual inspection, as well as paying annual fees and dues. Consequently, many small producers who generally follow the tenets above don't pursue organic certification, finding it too onerous and expensive. If strict adherence to the certification standards is important to you, you can certainly find whole animals to buy that meet these criteria. Or, armed with the list above, you might determine that your grower is close enough to the standards. For example, one producer I talked with was not certified organic but raised his cows exclusively on grass pasture that was not amended in any way. His operation lacked the logo, but not the spirit, of the law.

Natural

This is a term that you see bandied about more often than the word "sustainable," and it can carry a multitude of meanings (Cousin Kate finds this term totally meaningless, which might be the same thing). With regard to beef, the USDA says this indicates a product that has not been substantially (or "fundamentally") altered from its natural state. No artificial ingredients of any kind, including artificial color, may be added, and the product is to be minimally processed. I found growers use the word in a variety of ways in marketing their meat, to indicate that they don't use chemical fertilizers on their pasture, for example, or that their animals are not given hormones or prophylactic antibiotics.

Certified Humane

This label means the producer adheres to the Humane Farm Animal Care program standards (certifiedhumane.org), which dictate conditions for rearing, handling, transporting, and slaughtering animals as verified by an independent third party. The basic tenets, formulated animal by animal, range from providing wholesome and nutritious feed to using considerate and humane slaughter methods to avoiding administering antibiotics except when needed to treat disease. Smaller producers may follow the guidelines without certification.

Halal and Kosher

Certain observant Muslims and Jews eat only foods prepared in accordance with their religious practices. *Halal* is roughly translated as "permissible" and in this case refers to meat that is acceptable for Muslims to consume. Pork is never halal, as the Qur'an forbids its consumption, and halal animals are never slaughtered in facilities where pigs are also killed. The blood of animals is also *haram*, or not acceptable for consumption. In order to be considered halal, meat must come from animals that are butchered in a very specific manner called *dhabiha*. The process requires that the animal be killed by one swift cut to the throat while God is praised. All blood must be drained from the carcass immediately. Animals killed in this manner may not be allowed to see other animals being slaughtered nor the blade being sharpened.

Jews likewise have specific dietary laws, and kosher meat is slaughtered in very similar ways. As with Muslims, only certain species of animals are permitted to be eaten, including cattle and sheep but never pigs. The animals must be slaughtered by a specially trained *shochet*, or ritual slaughterer, in a certified facility. Meat and dairy are never allowed to come into contact with each other. As with halal meat, the blood must be drained completely from the carcass following slaughter.

What Kind of Cuts Can I Order?

Having determined the rough total poundage of meat you will be carting home to your freezer, it's time to determine which cuts will make up your order. Though it may sound daunting, the process is fairly simple. Every carcass is divided up into large sections called "primals." These primals are further subdivided into retail cuts. Hence, the processor carves out steaks, roasts, stew meat, chops, spareribs, and the like from each appropriate primal as instructed by you. Each animal-specific section of this book will tell you what the primals are for that animal and list the major cuts that can be gleaned from each primal.

A warning: When I went about ordering my own animals, I found my desires for specific retail cuts hemmed in not by a lack of information or by the grower, but by the processor. He (or she) who cuts your meat is close to God in this scenario, and sometimes God wants to cut your carcass the same way, or close to the same way, that he (or she) cut all the other ones that week.

Now, please hear me, I don't mean to dissuade you from pursuing the cuts you will use and want—this is a precious investment, after all. I simply wish I had understood from the outset the difference between what is ideal and what is practical. This is especially true when highfalutin' urban consumers who wish to cure their own *guanciale* (cured pig cheek) meet the more conventional processors used to cutting carcasses in the manner that the vast majority of consumers desire. As my friend Ethan likes to say, urban foodies can talk about nose-to-tail eating all they want, but 90 percent of the people going to restaurants order double-cut pork chops and filet mignon. Keep this in mind as you launch your own dialogue between you, the producer, and the processor. Here are some tips that may expedite the process:

» If you wish to make a special order or deviate from standard choices given on a cut sheet, do so in an informed manner and use the most universal and technical terms possible. Don't ask for "ribs cut Korean-style," for example, or "*kalbi* ribs." If you buy your beef from a locale where soy sauce is still sold in the "Oriental" section of the supermarket, your processor might not be acquainted with the term or the dish. Ask for "flanken-style ribs cut ¼-inch thick" instead, then share your recipe if your processor asks why you want them cut that way.

» Give the producer and processor plenty of lead time if you have special requests. If you want the tongue, for example, make that wish clear to both producer and processor before that cow goes anywhere near the abattoir (see the cautionary tale of the pig cheeks in Myth #2).

» If giving cutting instructions still scares you, ask the producer to guide you in making requests that reflect the way you cook. If your family is pot-roast-and-short-ribs-crazy, ask for the most cuts suitable for braising. Want to spend a lot of time at the grill? Ask for steaks wherever possible. If you want to make sausage, go for more ground meat, and ask for the fat. More flavor and less waste dictate bone-in wherever possible, whereas ease of carving might mean boneless at all costs. Or go for maximum variety: Once, when I ordered a whole lamb, I asked for each side to be cut completely differently so I could experiment.

How Much Will I Get and What Will It Cost?

The live weight of an animal is just that, the amount Uncle Dave's cow, for example, would weigh if you led her onto a large scale. When you take your dog to the vet, you take his live weight. However, when you're buying an animal to eat, most producers and processors charge per pound based on the carcass, or hanging weight, which is the weight of the animal after slaughter and usually with the blood, viscera, head, feet, and hide removed, depending on the animal (a cow head can weigh 25 pounds, so it makes a difference). Think of a whole carcass hanging in the cooler ready to be trimmed, like the sides of beef in a Mob movie.

To complicate things further, some producers combine the costs of the meat, the slaughter, and the cut-and-wrap into one cost per pound, while other producers charge a fee just for the meat and you must calculate an additional per-pound fee for the cut-and-wrap along with a set fee for slaughtering the animal. Does the math seem hard, Barbie? It *is* a little confusing. If this helps, here's the basic formula for inclusive pricing that includes all processing fees:

average hanging weight x inclusive price per pound = total cost

If you need to pay your processor separately, run this equation twice, once with the producer's cost per pound and once with the processor's, and then add their totals.

The final amount of meat you take home will be considerably less than the amount of potential meat you saw grazing in the pasture. We've already gone over the weight lost to slaughter, when the blood, viscera, head, feet, and hide are removed; once the carcass is at its full hanging weight, trimming and cutting the meat will result in further loss, such as if you order boneless roasts, or lots of excess fat is trimmed from the meat. There is also the weight lost in dry aging (see the sidebar "A Note on Dry-Aged Beef," in Chapter 6, "Cow") if you order beef. The final amount of meat you take home, in pounds, will add up to a percentage of the hanging weight sometimes called the cutout weight or finished cut weight, while the percentage of the hanging weight that translates into meat is called the carcass cutting yield.

I wish I could tell you that your animal will result in exactly *x* pounds of meat in your freezer, but the beauty of ordering your own animal is in the variability, right? How much meat you ultimately take home is the result of a complex algorithm that takes into account many factors, including how big your animal was at the time of slaughter, whether your animal is lean or fatty, well muscled or less so, whether you order bone-in or boneless cuts, how lean you require your ground meat, whether you take some of the offal or fat, and, in the case of beef, how long your meat is aged.

All of that said, I know you'd like a rough idea (you're one of those Detail People, aren't you?). Lucky you—in each animal-specific section I provide real-world examples of both costs and weights to help you in your planning process.

Who Will Kill My Animal, and How?

Well, if you're not going to do it, then someone else is—you know that, right? For some people, like my dear friend Ingrid, the less she knows about this part, the better. Though she was happy to purchase half a lamb, she squealed like a little girl when I was researching slaughter methods while sitting next to her in the office. Other people, like me, want to know exactly how this happens, because with knowledge comes some peace of mind. We all probably know of at least one horrific story in which an animal was butchered badly, or painfully, or slowly, or all of the above. (Cousin Kate has a childhood memory involving a slow-to-die pig that would have landed a more delicate lady on the therapist's couch.) I always think of those stories where people come out of anesthesia while they're having surgery. Wouldn't you do everything you could to ensure that didn't happen? Suffice it to say, if you are in Ingrid's camp, skip to the next section now. If you're in mine, following is some information for you to consider.

Some animals are dispatched by first rendering them unconscious or stunning them. This may be done with either an electric stunner that sends the animal into cardiac arrest or with what is called a captive-bolt stunner, which is a metal sleeve that holds a steel bolt propelled by compressed air against the head of the animal. The animals are then quickly bled out, a process in which the blood is drained from their bodies by "sticking" them, or cutting through the jugular vein and carotid artery in their throats. Animals are often hoisted up in the air by their hind feet, so that gravity can help drain the blood. This is a necessary step done to preserve the quality of the meat. Alternatively, animals may be shot in the head and killed before sticking, though the most complete bleed-out happens when the heart is still beating and can assist in pumping the blood out of the body. This requires either killing the animal simply by sticking, or by stunning and then sticking.

Many feel that it is not humane to string up a live animal by its hind feet, an uncomfortable position at any rate, and then cut its throat while it is struggling, upside down. I talked to one grower who did not feel it was humane to kill an animal by cutting its throat *period* and had worked with a local processor to ensure he upheld the slaughter methods he required. In this case, it meant animals never going to slaughter alone but only in groups with their herd mates, traveling to the processor enough in advance to acclimate to their new surroundings, being stunned before sticking, and being given both food and water in the 24 hours prior to slaughter. Many processors withhold food in the 24–48 hours prior to butchering in order to lessen the material in the animal's digestive tract and reduce chances of cross-contamination. But water should always be provided in order to prevent dehydration.

As with marketing terms, I was not able to determine *the* slaughter method that I could advocate for. In contrast to the grower described above, I spoke with another grower who thought the best kill method depended on the animal: A pig requires stunning or shooting because pigs are so sensitive to being restrained, while lambs will actually relax when held closely and tightly before having their throats swiftly cut. Having not done either, I don't feel qualified to personally weigh in. In fact, if I were to be totally honest with you, as a girl I found it difficult to whack the heads of the trout I caught on opening day with my dad, and those are *fish*. Small ones. If you want to know more about various methods and humane slaughter in general, peruse the collected works of Dr. Temple Grandin, an animal science professor who revolutionized the design of slaughter facilities and methods of slaughter with animal welfare in mind. It's fascinating stuff, and she offers hard science to back up her assertions.

Some parting thoughts for consideration:

If an animal is butchered according to either Islamic or Jewish law in order for the meat to be rendered halal or kosher, respectively, the animal may not be stunned or killed in any manner other than by having its throat cut with one deft motion. The reason given by both religions is to lessen pain and suffering on the part of the animal. Temple Grandin even has recommendations on best practices related to ritual slaughter.

Growers who have given thought to their slaughter methods and will speak openly and honestly with you about them probably engage in the methods they find most effective, if not also the most humane. Depending on your definition, they may be one and the same.

Even if the potential pain or fear on the animal's part is not paramount in the decision-making process, efficient slaughter is just part of good animal husbandry. Animals that are not bled out properly, that become overheated or stressed prior to slaughter, and/or that are banged about or treated roughly prior to being killed produce lower-quality meat, pure and simple. The muscles can be bruised through rough handling, and excitement or stress causes blood to travel to the tiniest capillaries out in the muscles, making it difficult to drain the meat thoroughly. Tiny hemorrhages and blood clots that remain in the meat are called "bloodsplash" and affect both appearance and taste. I know, I know—this is starting to sound less like food and more like an episode of *CSI*. Let me sum it up by saying that it's better for everyone involved if the killing goes quickly, goes well, and is performed by someone with experience.

That's the how. Then there's the who:

The grower might kill and bleed out the animal, then transport it to a processor. Some argue that it is more humane to kill animals at the farm where they lived, because then they aren't stressed in transit and are handled by people they know, Uncle Dave among them. Then again, professional processors might be more proficient at humane slaughter than the growers. Also, the bleed-out must happen quickly and the carcass maintained at the right temperature for best quality.

An approved mobile slaughter unit (or MSU) might go directly to the farm, so the killing and processing can take place there, again eliminating the stress placed on an animal in transit. Benefits of mobile units include greater custom killing and butchering options and greater flexibility, but some people argue that mobile units don't have appropriate oversight and therefore do sloppy butchering, or may not be USDA approved. Nevertheless, because the butchering comes to them, this is an option used frequently by smaller growers.

More typically, a live animal might be transported to a slaughter facility and killed and processed there. Note that slaughter methods vary *widely* from facility to facility. If a certain aspect of the slaughter process is important to you, ask your producer who will kill and process the animal and then contact them directly. You can also ask if the butcher is USDA certified.

Reflections on Butchery

○○○○○○○○○ ○○○○○○○○○ ○○○○○○○○○

As I sailed on a ferry toward Vashon Island, I (yes) tweeted that I was having *the* Seattle moment. It was a simply glorious morning. Mount Rainier handsomely loomed to the south while the fir-fringed shoreline tidily edged the city, hemming in the Port's industrial orange cranes and downtown's sleek towers jutting through the mist. I was signed up for an artisanal butchery and charcuterie class on Vashon, an island suburb that even among Seattleites is known as a hippie haven full of small farms and former urbanites who fled to find peace on a piece of land. It's not so different from any rural place on the coast, really, just colored blue on the political map rather than red.

I sipped my coffee, eagerly anticipating parting out my own pig. Carving up your own animal is quite different from looking at an exploded schematic on paper, especially for people with brains like mine for whom diagrams confuse rather than communicate. Knowing I was probably in store for a morning just this side of a guest spot on an episode of *Portlandia*, I couldn't help but think of Cousin Kate and wish she could be there, too. Like seeing a wild animal in its natural environs, it might help explain some of my nature to her and her people.

Entering the restaurant and butcher shop, I was not disappointed. Standing at the steel table was a handsome hipster with requisite nearly Amish facial hair, tying on his old-timey suede apron with the solemnity of a priest donning his vestments. The whole scene was rather peaceful: the butcher sharpening his picturesque carbon knives with a steel, the pig laid out on the table, and the rest of us murmuring in hushed tones as we waited for the class to start. I have to admit I was disappointed to see that there was only one pig—clearly this was to be a demonstration class rather than a participatory one, but I guess that would be a lot of pork for us to potentially ruin with amateur hack jobs.

For a dead animal, this was one serene-looking specimen. The farmer said it was a cross-breed pig that had been fed the waste from the organic yogurt company down the road, along with scraps from their farm. The head had been removed from the body and laid out on the table like we were going to do an autopsy. The eyes were respectfully closed. I should say the eye; the head was sawn in half lengthwise so we could see where the bullet hole went into the thinnest part of the skull, the skilled kill shot. We examined the plowlike structure of the head, and looked at which muscles would be used frequently by the animal while it rooted, or not much at all. I soon got over not having my own carcass to dissect.

It was a revelation to go through the animal muscle by muscle and see where the sirloin came from, to know where the picnic and the Boston butt divided on the body. I made up quirky little mnemonics to remember what I learned: "the pig trots over to the picnic"—i.e., the picnic roast is on the lower half of the shoulder adjacent to the leg, not on the top of the animal. That's the Boston butt, just like "Boston is at the top of the country" (relative to South Carolina).

"ventilate my hams"

Not even needing to channel Cousin Kate, I found some of the session a little over the top; I probably didn't need him to describe the trotter's "awesome richness" one more time—or to use the word "awesome" maybe ever again. A self-described "meatsmith" and practitioner of seam butchery, he practiced his art the *old* old-fashioned way, eschewing even the use of saws. He decried freezing meat *period*, preferring either to eat it fresh or cure it, and while I like the idea of a curing haunch hanging in the salt air, none of this translated as "sustainable" for my urban family unit if I wanted access to a whole pig.

However, the butcher was clearly a personable fellow, an artisanal processor who travels to farms and gives classes on processing animals. When he butchers a pig, he gives it a bowl of slop to munch and keep it busy, then shoots it with a .22, not right between the eyes, but on center and up a bit, where the skull thins out. He then sticks the pig to bleed it out right after it drops. I couldn't argue with the fact that if I were a pig, I wouldn't so much mind this guy killing me. He had reverence. He respected the animal before he killed it and he respected it after, going carefully over the body and not wasting a scrap. Just because I'm not a huge fan of blood sausage doesn't mean I can't appreciate his harvesting the blood and using it. Whether you come at it from a culinary point of view or a frugality angle, the end result is the same. Most of the pig was being used, and it would surely taste delicious.

I left enjoying the romance of the story that he deftly wove around this pig, every pig, the process and the interconnectedness. I would happily nominate this guy to dispatch every animal slaughtered in the region. I wasn't surprised to learn later he was a former student of Renaissance literature.

But clearly his methods didn't translate directly to my busy city life. During the session, one student who gamely cures meat in a box in his garage asked our instructor about the composition of his own curing chamber. "Oh, I just hang my meat near a window that I leave open a little year-round," he answered. Our moist sea air and mild climate make curing easy, he said. I snorted a bit, imagining what and who might come through my South Seattle window should I leave it cracked year-round to ventilate my hams. (It would turn out that what was *inside* my house would prove to be more of an issue. Note to future charcutiers: Keep your sausages well out of reach of your tallest dog.)

At the same time, I loved the simplicity of it. Using buttermilk to jump-start the good bacteria in your salami instead of measuring out spoons full of pink salts, and talking about hygrometer measurements. Waiting for *two years* while my handmade prosciutto turned buttery and rich. But what was required to facilitate this way of life was to me, in its own way, as difficult to parse as the instructions for building a professional curing chamber at home. City vs. country life, organic vs. conventional, foodie vs. working mother, big vs. small...the day hadn't answered any questions definitively for me, but I definitely had more food for thought.

Where Will I Pick Up My Meat?

There are several options for how to obtain your meat once it is ready (which may include further processing like smoking or curing or aging in the case of beef).

From the processor: The most direct means of getting your meat is trekking out to the processor once it is ready. If you paid the producer separately from the processor, this is probably what you will be asked to do. Before you buy a beef from a producer, ask where the animal will be processed, or broken down into cuts and wrapped. The producer might not be so far outside of town, but the processor could be another half-hour away, in the other direction. If you don't have a car, you'll need to bribe a friend with a steak or two. Driving all over hell's half-acre isn't particularly good for the planet, or your busy lifestyle, but it can be the cheapest option.

From the farmers market: If you met your grower at the farmers market and you ordered only a quarter-beef or less, you may be able to pick it up there on market day. More than a quarter and you will probably need to go to the farm or to the processor. Smaller animals like goats or lambs are easier to transport and are more likely to be available for market pick-up.

From a drop location: Lucky for us urban folk, some of the nice growers who raise meat outside the city now drive refrigerated trucks into the city loaded with our meat. On an appointed day, you travel to a predetermined location and pack up your meat, usually for a nominal fee. This is my preferred method. You still get to have an in-person exchange with the grower, there aren't as many people as at the farmers market, and it involves the least driving. Some meat producers use pre-established in-city locations that other farmers use for community supported agriculture (CSA) drops.

From the mail: Okay, make that last option the one that involves the *second*-least amount of driving. Some producers will indeed ship your meat to you. It ranges from expensive to very expensive, but if you're already spending your monthly paycheck on meat, maybe a few hundred for shipping doesn't bother you. Make sure you're home to receive it; no one wants that meat to spoil on your doorstep when you've decided to go mountain biking.

The Miller Ice Chest
(*Or*, How to Transport Your Meat)

○ ○ ○ ○ ○ ○ ○ ○ ○ ○ ○ ○ ○ ○ ○ ○ ○ ○ ○ ○ ○ ○ ○ ○ ○ ○

Husband Hendel loves the trope of "the Miller Family." My family of origin is his favorite running gag, a clan he paints as gun-toting, bird-hunting, dog-loving, loud-mouthed epicureans who can't have civil political discussions or cook for less than twenty. "Well, the Millers will be cooking," he responds to questions about our holiday plans, meaning my brothers and I will squabble good-naturedly over dominance in the kitchen, trying to one-up each other as we churn out far too many dishes than either the available counter space or the bodies in attendance could possibly hold. But sometimes big is good.

When I checked into the ice chest situation prior to picking up my first pig, I found it woefully inadequate. The drive would not be short, and I had an entire pig to transport. Thanks to the Millers, I knew where to turn. You see, every year as I grew up, my brothers and dad would eagerly await the arrival of the inch-thick holiday edition of the Cabela's catalog so they could make up their Christmas lists. Waders, camouflage for every season, specialty gloves, ammo— it was a sportsman's dream come true. And as luck would have it, years later a gigantic Cabela's superstore was built right off Interstate 5, an hour from Seattle en route to my pig. I cajoled my colleague Ingrid into going with me, promising we'd talk about a shared client on the way home and then stop for diner food in a small town out in The Country. And just one small pit stop, I told her: We'd have to stop in for a cooler on the way.

The store has its own freeway exit, which becomes a twisting road eventually leading to a manicured parking lot and an entrance framed with stuffed animals. We stepped inside, and Ingrid grabbed my arm as we drank in two full floors of archery equipment, camping gear, fishing rods, deer blinds, pocket knives, and of course many, many guns. In the center of the building was a faux mountain on which perched all the things one might want to shoot: bears, mountain goats, deer. Little pheasants roosted on the side. "It's like REI, for *different people*," she whispered to me, referencing that little Seattle co-op-turned-outdoor-gear-behemoth Recreational Equipment Incorporated.

We found our way upstairs, me glancing lovingly at the smokers and food preservation section, until we faced the full selection of coolers. I wrestled them down from the shelves and opened them one by one like a little country Goldilocks. "Not big enough." "No wheels." "Too expensive." Finally, opening the lid of a wheeled, 100-quart Coleman, I told Ingrid, "This will do."

And do it did, holding one whole pig with only two fresh ham roasts and the back fat having to ride shotgun. After I got home and unpacked the pork, I set the cooler in our basement alongside our other sporty gear. Husband Hendel came in and spotted it. "Oooh," he cooed approvingly, inspecting it like a new car. "It's Miller-sized."

I have found my ice chest incredibly useful for all sorts of tasks outside of hauling around carcasses, but rest assured you do not need to purchase an ice chest, let alone a Miller-sized one, in order to pick up your animal. When you collect your meat it will be *frozen*, as in rock-hard, could-be-used-as-a-weapon frozen. It will easily last an hour or two simply riding in a cardboard box, especially packed in with all the other frozen meat to keep it chilled. Make sure to bring your own box, however—or two, or three, or four—when you go to pick up your animal. Processors usually don't provide them.

MYTH #4

If I find out the truth about where my meat comes from, I will become a vegetarian!

You might. That's how it worked for Jonathan Safran Foer, author of *Eating Animals*, and many others who somehow didn't feel the same need to tell us about it in excruciating detail. But let me say a few words here as someone who was a vegetarian for more than a decade.

Though people who know me now find this very hard to believe, from the time I was a senior in high school until after I married Husband Hendel I was a lacto-ovo vegetarian who had a brief fish period and sometimes ate mollusks and sometimes just didn't eat anything that "had a face." I was regularly ridiculed by my family, needless to say. I had plastered on the door of my dorm room one of those photos of veal calves in a little cage. You get the picture. And much of what I thought was disgusting then about how some meat is raised I still find objectionable. But what I've found is that the more I learn about the animals—who raised them and how they were fed and bred and even how they were killed—the better I feel about eating this kind of meat, this way. My food journey has made me actively question anonymous meat, as it did my children. Husband Hendel is less bothered. I still eat meat in restaurants, even in some where the Health Department would have a field day. But I actually eat less meat out than I used to, and less meat in general, because at home I eat meat I *know*.

Like making babies, it is not really a mystery how meat is produced. However, there is a lot of variation in what happens between birth and death for the animal. If you're going to eat dead things, own it: Figure out what works for you, make your peace with it, and accept those standards as your own values. Don't rely on politics or national marketing associations or me or your three-year-old's classmate's parents to make this decision for you. And if you are still turned off by growers who care about their animals, raise them the best way they know how, and try to produce good food for you to eat, you probably shouldn't make meat part of your diet. Period.

5

HOW DO YOU
KEEP AND USE YOUR
WHOLE ANIMAL?

Buying meat by the whole animal is not like shopping at a store. You can't decide in the morning that you'd like half a beef and start cooking it that afternoon. And that's a good thing, because a whole animal is a purchase that deserves advanced planning (trust me, I know), and a mandatory waiting period is built in to save you from yourself. Depending on the size of the producer and the type of animal, you may even need to wait months before your animal reaches market weight or until slaughter time. Lambs born in the spring, for example, are usually not ready until early fall. Some farms process pork only when the weather is cooler. Animals raised exclusively on forage take longer to reach market weight than those fed grain and more calorie-rich foods.

Once your animal is slaughtered, you may have to wait for your beef to age or for your pork cuts to be smoked or cured. I have listed in each animal section the approximate time you can expect to wait before getting meat in hand. In general, think of this less as a sprint than a marathon. Take time to buy the animal you want from the grower you respect, and then enjoy having the meat at the ready for months to come.

Freezing

In general, well-wrapped frozen beef lasts nine months to a year, while pork, goat, and lamb should be used within six months. Some people say that vacuum-packed meat lasts longer than meat wrapped in butcher and freezer paper, though it does not look as old-timey and picturesque. Others find that vacuum-sealed packages are more delicate and can lose their seal if moved around too roughly. Your meat will not go bad so long as it stays frozen, but you will notice a decline in its texture and taste as time goes on. Poorly wrapped meat or meat that has partially thawed and then refrozen is in danger of freezer burn. Any grayish patches should be removed before cooking. If the discoloration is extensive or the meat is covered in ice crystals when you unwrap it, consider throwing it away, as the taste and texture will surely have suffered.

Do you need an additional freezer? Fifty pounds of meat will fit in 2.25 cubic feet of freezer space, adjusting for the size of packages and layout of the space. If you want to invest in a freestanding freezer there is additional cost, but the freezer will keep your meat 5 to 10 degrees colder than your refrigerator's freezer compartment.

What freezer should you buy? Looking at energy usage, efficiency, and overall cost, you should consider the following points:

» Chest freezers are more efficient than uprights. They are better insulated, and not as much cold air escapes when the door is opened.
» Chest freezers, in my experience, are harder to keep well-organized if your meat turnover isn't high. What sits at the bottom languishes at the bottom; out of sight, out of mind. Cousin Kate uses her chest freezer to its best advantage, but she's more organized than I am in general.
» Automatic-defrost freezers can use up to 40 percent more energy than manual-defrost freezers. However, if you're like me and will realistically never defrost your freezer until you move, then it's nearly a wash, as you must regularly defrost freezers for them to work at full capacity.
» There is an Energy Star database you can use to compare reported efficiencies of different models. Some locales offer rebates for homeowners

who trade in appliances for more efficient ones. Check with your city's utilities company or environmental department for more information.

Buy a freezer just big enough for your needs, and place it in the coolest part of your home or garage. Make sure air can circulate behind it. To maximize efficiency, freezers should be kept full, with just enough room for air to circulate. If yours starts to empty out, it's time to buy another animal!

Locker Meat

○ ○ ○ ○ ○ ○ ○ ○ ○ ○ ○ ○ ○ ○ ○ ○ ○ ○ ○ ○ ○ ○ ○ ○ ○

When I mention "locker meat" to people, I find they either get the reference immediately or they don't have the foggiest idea what I'm talking about. I thought of this recently when I saw a sign by the side of the freeway advertising LOCKER BEEF in big bold letters.

I remembered driving out to Cliff's Meats, essentially a butcher shop and storage facility a few miles outside of town, when I was a kid. My mom has told me that back then many folks depended upon a meat locker to get dinner on the table. We had a key to our very own locker, which we could access whenever Cliff's was open. Inside was not only frozen meat we bought from Cliff's—they processed animals and did the cut-and-wrap, made sausage, etc.,—but also meat we didn't have room for in our freezer at home, such as elk, duck, and pheasant from hunting.

Mom said even Grams, who is 96 and still lives on the farm, had a meat locker "in town," because if you slaughtered a cow, *and* shot an elk, *and* butchered a pig, that was a heck of a lot of meat, and no one had their own freezers then. Meat lockers were both readily available and frequently used by folks in the neighborhood. "Back then we didn't move around so much like people do now," Mom said. "Our neighborhood never changed, you know. Back then you got a job and you worked at one company until you retired. It made it easier to keep a lot of meat in one place and know it would be convenient."

The place I spotted that sold locker beef hadn't changed much, I suspect, since the decade when my mother was a young wife in her stable suburban neighborhood. In Seattle, however, CSAs and the renaissance of neighborhood butcher shops, along with a general reduction in meat on hand, have made locker meat a fairly moot concept. My local butcher shop, Bob's in Columbia City, sells meat "bundles" or assortments of 25–75 pounds of beef, pork, and chicken, but they don't have the room to store your meat after you buy it. Attention hipsters and foodies: I see a small business opportunity here.

Thawing

Chest freezers and standing freezers have the advantage of keeping meat colder than the freezer compartment of a normal kitchen refrigerator. That's the upside. The downside is that they keep meat colder than the freezer compartment of a normal kitchen refrigerator. This means you have to plan, and I mean really plan. Large roasts or whole legs or shoulders can take *days* to thaw in the refrigerator—I've had some take as many as five. Smaller and thinner cuts thaw more quickly, as does ground meat, which thaws perhaps the fastest. Once thawed, you have two to three days to cook it.

"But I want dinner *now*," you say? Well, there a few ways to go about it.

» **Metal thaws meat more quickly.** It's an old trick, but a good one. Remove the meat from its wrapping and place it on a sheet pan on the counter. Turn it frequently. This works especially well for cuts like chops or steaks, and it even speeds things up in the fridge. One word of warning: Never salt your meat before placing it on an aluminum tray; aluminum reacts and can cause off-flavors.

» **Brine or marinate the meat in a frozen state.** Though you may find your whole-animal meat, especially pork, doesn't need brining, brining still works if you want to add flavor or prep for smoking, for example. It's okay to place the meat cuts in the brine or marinade frozen; just allow for extra brining or marinating time.

» **Begin cooking the meat in a frozen state.** Bacon is super-easy to pry off and begin to fry. Ground meat that needs to be browned doesn't suffer if it's frozen. Though many consider this completely verboten, and it's certainly not ideal, in a pinch I've started braised roasts for soups with frozen meat, when caramelizing the exterior wasn't needed. It's an easy choice if you start dinner on a weekend morning.

» **Use cold water.** Here the advantage of vacuum-packed meat (Note: Only my goats have come this way) becomes clear. Plop the cuts, package and all, into a bowl and fill the bowl with cool water. Either change out the water every half-hour or set a faucet to drip into the bowl. If your meat's not vacuum packed, you can remove it from its butcher paper and place it in a resealable bag, then continue as above, separating chops or steaks if bundled together so they thaw more quickly. I don't like setting meat in cold water directly. Once thawed, this meat needs to be cooked right away and may not be refrozen.

» **Use a microwave.** Do this attentively and turn the meat frequently. Even if you're very careful, however, the meat will inevitably heat up around the edges or in hot spots. Be prepared to cook it immediately if you use the microwave to defrost—there is no turning back.

Cooking

Before you get into the nuts and bolts of the recipes, I need to tell you something about my personal limits. Fergus Henderson, Britain's evangelist of nose-to-tail eating, can eat all the spleens he wants. I wish I were that person, but I'm not. In other words, this is not the place where I out-foodie you by telling you how much I love to drink the blood of the pig as I stick it. This is where I say that I've had some brilliant tongue and I'm a sucker for sweetbreads. I love crispy gizzards and find pig feet succulent, if a bit rich. Rillettes—shredded pork poached in its own fat—are delicious, in small quantities, with cornichon and mustard, but my kids aren't making any moves to polish off the rest. Kidneys are lower on my list—heart, too. Heart is grainy—put it in your pepperoni, sure, but don't center it on my plate. Liver's okay, but not fried with onions. I am not a big fan of tripe. I've had it, multiple times; I don't want to eat it. I want *someone* to eat it, just not me.

When my friend Ethan surprised me with sixty Shigoku oysters last Christmas, I gave him frozen lamb and goat liver, kidney, and heart. (Not only nutritious, they make great stocking stuffers, too!) But Ethan just isn't like everybody else: He cooks dishes like "Party Tripe." If you're like me, get a friend like that in your life when you put in your order for your animal. Or get out your sausage grinder.

That confession aside, if one is buying a whole or partial animal, one should certainly share in the offal, or "variety meats" as they're called, the bones, the fat, the fun. If you want to cook the pig head you're in for a treat, but I'm not here to guilt you into making your own haggis.

For me, this started as an adventure and grew into a way of eating that made me feel good. And along the way, something curious happened. I started making jam. I picked tomatoes and dried them in the oven. I rendered huge pots of lard and made delicious pastry with it and *then* made pies, pies full of blackberries that I picked before pulling the canes out of my rosebushes.

To be honest, it scared me a little. It was like living with my fixie-riding Portland doppelgänger. I nearly knitted. And although I joyfully made fun of myself while I was doing it, I also recognized that buying my lambs and goats and parting out the big animals and figuring out how to cook them in diverse and delicious ways had made me both thriftier and more home-spun, in the same way that someone who installs her own light fixture feels suddenly handy.

You see, I grew up in a house with a "fruit room," the place where canned peaches floated in cloudy syrup and jars of pickles stood at attention. I ate homemade peach leather dried in the Yakima sun and dehydrated pears picked off our trees. *Preserving* held no artisanal cachet for me, new as it was for many of my friends. But Uncle Dave's cow led me somewhere new, to a place where I received a pure kind of joy from taking the food growing and living around me and using it to its fullest and most delicious extent. I could always cook, but this was less mastering something than letting go. It wasn't "local" or "sustainable" or "green." It wasn't a buzzword. It was just good.

A Note on This Book's Recipes

I'll assume that you can sort out how to do a basic braise, or sear a chop, or even make sausage–cornbread stuffing. But not everyone has a stock family recipe for whole fresh hams, or beef cheeks, or oxtail. And for those of you who do, maybe you've never made a proper Trinidadian goat curry, or lamb kibbeh, or you'd like to try roasting a whole goat leg. In putting together the recipes included in each of the following animal-specific sections, my goal was to come up with fun ideas for basic parts and solid ideas for fun parts (within reason), all truth-tested by that non-discrete-meat-eating Husband and those two sauce-skeptical children, along with a community of meat-loving friends and family. If you're an urban foodie, now you can make some of your global take-out faves with your own animal at home. When you're through with the recipes, you may be ready to bring some goat to your next potluck, or pack homemade jerky in your kid's lunch instead of a Clif Bar. Enjoy your meat!

PrE-agricuLTural FARMers MARKET

A Sign Things Have Changed

○ ○ ○ ○ ○ ○ ○ ○ ⬭ ○ ○ ○ ○ ○ ○ ⬭ ○ ○ ○ ○ ○ ○

Oh, it was a busy night. I was packing up our kitchen for the impending remodel and trying to empty out both fridge and freezer. One child had been diagnosed with "parvovirus B19" (a rash) and sent home from school that morning. It was nearly Thanksgiving, and there were trips to plan and the house to prepare, and work was busier than ever for me. We barely saw Husband Hendel. There was scarcely time to remember to feed the children; they found me when they were hungry, and I discovered hot dogs in the meat drawer and served them. (I cooked a green vegetable as well, I hasten to add.)

When Atticus tried to get my attention, I was so preoccupied it took a minute to understand what he was asking me.

"Is this ours?" he asked a second time, hot dog in hand. "Did you make it?"

I still wasn't quite sure what he meant. Yes, of course it was for him and yes, I had made it.

"Yes, I...wait...*ours?*" I asked.

"Is this from one of our animals?" he clarified.

Ah, the guilt! Not only was I feeding him a hot dog, but he had connected the dots, clever child. He had gone with me to the butcher that day to pick up additional fat for the sausage I would be making with his uncles at Thanksgiving—sausage from our pig downstairs, sausage I planned to cure in the basement—and he knew that *this* meal was less cooked for him than merely heated. I was a terrible mother.

"No."

I took the time to sit down and look him in the eye. "It was made by a Seattle company, though," I added. It was true—the dogs had come from a meat shop in Pike Place Market.

And that's when it hit me. This wasn't a moment to feel guilty about; it was a moment to feel proud. No, not every bit of meat that hit their plates would be perfect, its provenance carefully charted. Not every meal would

hot dawg!

be complete, exhibiting artistry and nutrition in equal measure. Hell, not every dinner would even include the four of us sitting down together, or the children wearing pants. But the journey from Uncle Dave's cow had made us all more aware of the question. Unbidden, Atticus had looked at his plate and seen not a hot dog, an icon, but more correctly a type of sausage. And he considered not only what sort of animal, but which specific animal it could have come from. More important, it was now clear I had a mandate to master a new sausage-making realm: Emulsion sausages, here I come.

6

COW

There is no better place for us to begin than with the animal from which this book takes its title. Yes, goat is a newcomer to this country's it-meat scene, and pedigreed pork's wave is just now cresting, but you can never go wrong flying your meat flag high with the classic American red. Beef is a verified crowd-pleaser that has never gone out of style. And though my first adventure with Uncle Dave's cow would lead me through a rotation of pigs, goats, and sheep, do you know what happened? By the time I'd come full circle, I found myself once again hankering for a delicious bit of beef.

As elegant as prime rib and as down-home as meatloaf, beef offers a great range of quick-cooking cuts and roasts well suited to Sunday dinner. Buy a whole beef and your family (and neighbors) will be in meat for a year.

Corn-fed Midwestern beef was the iconic meat of yesterday, but if you buy a whole or partial beef today you can make many more choices regarding your animal's diet, from organic grain to pure grass pasture. Grass-fed beef's nutrient profile is touted by producers and health journals alike, lower in saturated feet and higher in the good stuff, like vitamins and omega-3s, but if you are used to corn-fed, I'd urge you to do a taste test before committing to a whole (or partial) grass-fed cow.

One University of Chicago researcher cautions that consumers should be prepared for the "distinct grass flavor and unique cooking qualities" of grass-fed beef, along with fat that exhibits "a yellowish appearance from the elevated carotenoid content (precursor to Vitamin A)," mostly stemming from fatty acid content differences from grain-fed beef. Though this may sound a bit like she's describing a smart girl as having a good personality, all it means is that grass-fed beef is much leaner and therefore requires shorter cooking times, that it has delicious fat tinted a bit yellow from the same sort of stuff that makes carrots orange, and that it has a more pronounced flavor. Some say it truly tastes like beef, I among them, while others can't cotton to the more minerally taste. Chicken-breast-no-skin kind of folks might need a ramp-up period, or may choose to buy grain-fed or grain-finished beef because of the differences in taste or its more tender texture.

What Kind of Cow Can I Get?

Grass-fed, organic, pastured…the terms are more dizzying and the options more varied if you are buying beef than for any other animal. Whether you are doing preliminary shopping for a beef producer via the Internet or chatting up a grower at your local farmers market, you'll be well served to know your terms before you begin. If you haven't read the earlier "Isn't *All* Meat 'Natural'? Meat Marketing Terms Demystified" section in Chapter 4, you should do so now. Then read the section that follows here, to see how those terms apply more specifically to beef. AN IMPORTANT NOTE: Before you go having conversations with growers about their cattle's specific diet, I should probably tell you that, contrary to the name of this book, one doesn't buy a whole or half cow; you buy a portion of a *beef*, a *quarter-beef*, and so on. But *Uncle Dave's Beef* just doesn't have the same ring to it, does it?

Grass-fed. Wait, don't all cows eat grass? Indeed they do, at least for some part of their lives after they are weaned from their mothers. But as I'm sure you're aware in this post-Pollan era, since the 1950s the majority of cattle produced in America go on to eat other things. These include corn, which, like the excess salt-and-vinegar potato chips eaten by a woman with PMS, can disturb the delicate balance of one's system and furthermore serve to pack on the pounds. In the cow's case, the latter effect is considered desirable. Some consumers insist on grain-fed beef because it is more tender and has a milder flavor. Grain-fed is the beef with which most of us are acquainted.

Before the USDA revised its definition, cows called "grass-fed" could have primarily eaten grass during their lifetime, then been put on grain for the last few months before slaughter. The grain-finishing would put weight on the animal quickly. But it would alter the nutritional profile as well as the taste of the meat, with one study concluding that regardless of breed or age or locale, what made the significant and most consistent differences in the overall fatty acid profile and antioxidant content of the meat was whether the cattle ate grass or grain.

In fact, a plethora of studies suggest that grass-fed cattle produce meat that is lower in saturated fat and total fat, and higher in cancer-fighting properties like conjugated linoleic acid (commonly called CLA). Exclusively grass-fed beefs have higher levels of cholesterol-neutral stearic acid and less myristic and palmitic fatty acids, both of which serve to elevate cholesterol levels. Grass-fed beef is higher in omega-3 fatty acids (and arguably tastier than fish oil capsules) as well as vitamin E, beta-carotene, thiamin, riboflavin, calcium, magnesium, and potassium. For buyers for whom nutrition is a main criterion in purchasing meat, knowing whether or not cattle maintain this profile until slaughter is important.

Most conventional growers finish cattle on grain because it causes more rapid weight gain than grass, as well as increasing the marbling in the meat. Grass-finished cattle are not fed grain at any time during this period, but instead only grass, hay, clover, and the like (see the USDA definition earlier in the book) until the big day arrives. Why is this important? What do a few months of grain do after all that pasture? It's up to you to decide what to believe, but articles in the *Journal of Animal Science* and human health journals alike suggest that many of the health benefits of a grass diet outlined above wane when cows are finished on grain. However, others hasten to point out that the health benefits of this "good" fat in such lean meat could be negligible. And flavor profiles even among grass-fed beef vary depending on the overall health of the animal.

If the health benefits and taste of grass-fed and grass-finished beef are important to you, know that you will pay extra for the privilege. Grass diets mean cattle take longer to come to slaughter weight, and that extra growing time is a cost borne by the grower that will be passed on to you. The easiest way to find out about the dietary habits of *your* beef in particular? Ask.

Should I Care What My Cow Eats?

Like growing people, growing cows is a more complicated business than you might think. Reap the benefits of dealing with an individual grower by asking him or her directly what your cow will have eaten in its lifetime, how the feed was grown, and why they use the feed they do. Other points to remember:

» Exclusively grass-fed meat is chewier, leaner, redder, and some say more "beefy" tasting than grain-fed meat.
» Grass-fed meat is generally reputed to be lower in saturated fat and cholesterol.
» Cattle breed as well as feed can affect marbling, texture, and taste (see below).
» If you can, try before you buy. Many producers sell smaller packages or bundles of meat that will let you sample the flavor and tenderness before committing to a whole animal.

What Breed of Cow Should I Choose?

Cattle breeds fall into one of two camps, dairy and beef. Beef cattle are bred for muscles and marbling and to gain weight fast, while dairy cows are bred to be prodigious producers of milk. (Even so, for all breeds milk production is a consideration; you want even beef cattle to be able to feed their offspring well.) In addition to milk, beef cows are bred for such factors as uniformity, disposition, fertility, weight, hardiness in different environments, and polling (polled cattle lack horns).

Unless you're Warren Buffett out shopping for a whole Kobe, you will probably be choosing from an Angus or Angus cross (reputed to have the best flavor) or a Hereford (the most common breed of beef cattle). The Brahman is a breed that hails from India, often crossed with Angus to produce the "Brangus." The Beefmaster, which sounds to me more like a grill or a seasoning salt, is actually a three-way American cross between Hereford, Shorthorn, and Brahman cattle. Less common are Frenchies like the Limousin or the Charolais, big, cream-colored cattle reputed to have leaner meat.

Just this much information is actually plenty for most urban consumers. Leave the cattle breeding to those who raise cattle. In order to keep you from embarrassing yourself, I will tell you that those iconic black-and-white spotted cows plastered all over your organic milk carton are Holsteins, and they are dairy cows. Dairy cows are slaughtered but are not usually marketed as whole beefs to consumers. Uncle Dave says they actually marble well and make very good meat, the situation being that the cut-out from hanging weight is less than with beef breeds.

What Cuts of Beef Can I Order?

When the book *Good Meat* first came out, I lovingly read what author Deborah Krasner wrote about the educated trade-offs one can make in terms of ordering cuts from each primal, and of how to persuade your processor to give you the strange, the cosmopolitan, the unique cuts. If you can find such a willing processor and wish to invest the time and energy studying and conveying your options, I would refer you to the excellent descriptions beginning on page 48 of her book. More likely, though, you will be hard pressed to get even the tongue to accompany your order of a whole or half-beef, much less having a traditional *bistecca fiorentina* cut carefully from the center of your short loin.

Some of the toughest cuts, scrap meat that is trimmed away to form the cuts you request, are combined with meat scraped from the bones and ground together to produce hamburger. When dealing with an entire beef, it's not just the volume of meat you receive that must be considered, but also the percentage that ends up as hamburger. About 40 percent of your total order, or 160 pounds from a whole beef, will likely be ground. Whether that amount gives you pause or has you celebrating (it's certainly versatile), your cutting instructions will affect to some extent how much beef is ground and which cuts end up in your freezer.

If you are ordering a quarter or eighth of a beef, know that your choices will necessarily be more limited. One does not order a *specific* quarter or eighth—that is, you don't just point to a part of the cow you like, like choosing the piece of cake with the most frosting. What you will receive is a portion of the total mixed cuts from the fore- and hindquarter of the animal. Given those constraints, there is only so much trading you can do. It is perhaps here, in the cutting of the meat, that you will reap the most benefit from going in with family or friends and ordering larger pieces from the producer. The larger the portion of the animal you are working with, the more flexibility you will have to get the cuts you want.

Primal Needs

As I said, every animal is divided into basic slaughtering areas or sections called "primals," which form the basis for the universal language of meat cutting across these United States. Pigs, lambs, and goats are divided into fewer subsections, but all kin of Uncle Dave's cow are divided into eight: Chuck, Rib, Short Loin, Sirloin, Brisket and Shank, Plate, Flank, and Round. Some of these terms probably ring a bell with you in culinary terms—"flank steak" perhaps, or "chuck roast"—while others may seem fairly clear anatomically. For example, I'd hope you could get close to the "rib" primal on a cow blindfolded (you, not the cow). There is more to the rib primal than ribs, however. So let's look at a basic primer on the primals, so you can make the most informed choices, or try to.

Beginning with the animal's shoulder, we have the **Chuck**. This is one of the larger primals, accounting for about 25 percent of the total carcass. It extends from the front of the animal back to include five rib bones, as well as some of the shoulder blade and arm bones. Think of a cow in action—the forelegs support 60 percent of the cow's weight—and you'll get a sense of what to expect from meat in this section. Think moist heat and slow cooking to break down the connective tissue contained in these cuts (*Hello, braise!*) and, as with other muscles that get a lot of use, you'll be rewarded with lots of flavor. Because of the toughness, the chuck is often ground or cut into generic "stew meat." One exception, called out by both Jason and Uncle Dave, is the flat iron steak. Cut from the top blade, it is reputed to be the tenderest cut on the animal after the tenderloin. It's a nontraditional cut, and you'll need to ask for it.

Beef Chuck Cuts:

Boneless/bone-in chuck steak   Arm roast

Boneless/bone-in chuck roast   Stew meat

Seven-bone pot roast   Short ribs

Top blade roast   Ground chuck

Top blade steak

Heading from the shoulder down the front leg, we come to the **Brisket and Shank**, which make up about 10 percent of the carcass. Most people are familiar with brisket, whether it's their Jewish grandmother's favorite slow-cooked recipe, pastrami or corned beef, or Texas BBQ–style spiced and smoked. Those arm bones are full of goodness and will make a rich stock faster than you can say *Holy cow*. The meat from the legs can be cut off and braised. (This is often called "beef shin" by Asian butchers and it's *delicious*.) If you'd like to try your hand at corned beef or pastrami, ask for the brisket to be divided into the *flat* and the *point*, the flat being the cut you'll use.

Beef Brisket and Shank Cuts:

Brisket

Crosscut arm bone

Beef "shin"

Back to the top, we come to the **Rib** primal, which makes up another 10 percent of the whole carcass. If your family is anything like mine was growing up, the most celebratory occasions called for an enormous prime rib roast, served bloody rare with whipped horseradish sauce. Cut that roast into steaks and you've got a ribeye.

Beef Rib Cuts:

Prime rib roast   Beef ribs

Boneless ribeye roast   Ribeye steak

I like to think that if you were to put a saddle on a steer, you'd cinch up the saddle right between the **Plate** and the **Flank**. (You could argue on the placement, and if you *have* saddled a steer—or bull—before, I will defer to your opinion.) Let's discuss these primals as a unit, since together they add another 10 percent to the carcass weight.

I've heard tell that flank steak used to be a cheap cut, not anymore. Maybe it's so expensive these days because we recognize that the flavor is deep and delicious, and that it's no bother to marinate it, cook it just to medium rare, and slice it thinly against the grain. You can do the same with skirt steak, the diaphragm muscle of the cow, which also hails from this region. Processors will generally slice that and call it "fajita meat," though it's a tasty cut on its own.

Beef Plate and Flank Cuts:

Flank steak	Fajita meat (sliced skirt)
Skirt steak	Ground

Okay, so the steer didn't like the saddle and is trying to buck you off. Right about where your arse would be bouncing up and down we have the **Short Loin**. Located just behind the ribs, the short loin makes up less than 10 percent of the total carcass weight on average. Some of the most traditionally choice steaks come from the short loin. There are choices to be made here. You can have the entire tenderloin removed and cut into steaks (filet mignon), or keep it whole and make the center into beef Wellington. I have made beef Wellington with this most luxurious of cuts, but I don't know if I'd rather eat beef Wellington or a T-bone steak. You can't have both: One side of that classic T consists of that tenderloin, while the other is a New York strip. The T, which makes sense once you think about it, is just a portion of the backbone. A Porterhouse is cut from farther back on the primal and has even more tenderloin included in each steak than a T-bone.

Beef Short Loin Cuts:

> T-bone steak
> Whole beef tenderloin or center-cut tenderloin
> Tenderloin steak (filet mignon)
> New York strip steak (aka boneless top loin steak)
> Porterhouse steak

Moving on back, we arrive at the **Sirloin.** If a cow stood upright, you might call this the small of its back. This diminutive primal offers not much more meat than the short loin. It is tender, flavorful meat, just not quite as tender as meat from the Short Loin. Excellent steaks and roasts come from this primal, including the increasingly popular tri-tip, which I had never heard of before a long-ago California boyfriend raved about it. This cut, taken from where the sirloin meets the flank and the round, goes by a host of monikers including the much fancier *culotte*, which is what it was called on the bistro menu when I ordered it with *frites* not two nights ago.

Beef Sirloin Cuts:

> Top sirloin steak   Round tip roast
> Sirloin tip roast   Tri-tip roast

For those who appreciate a nice, **Round** rump, the location of this primal should be easy to remember. The round encompasses not only the rump but the entire hind leg of the animal, making this the biggest primal of all. Meat in this primal is generally flavorful but lean, and can include big, full-cut round steaks—sometimes called London Broil or Swiss Steak—or bottom, eye, or top round steaks instead. You can ask for parts of the round to be run through a mechanical tenderizer, which results in a cube steak or minute steak, though I think anyone who calls this cut anything but Chicken Fried Steak (prepared accordingly) is insane. Unless you've picked up a jerky slicer from Cabela's to do the job, you could also ask that a portion of the round be sliced super-thin for that purpose. Good roasts come from the round, as well as shank, which is delicious for braising.

Beef Round Cuts:

Full-cut round steak   Rump roast
Top round steak   Cube or minute steak
Bottom round steak   "Kabob meat"
Eye roast   Hind shank

Working with Packaged Cuts

Especially if you order a smaller portion of a beef, you may be offered a "package" or a list of suggested cuts. This is certainly easier for the processor, who may then cut all carcasses from the same grower the same way. It also provides consistency for the grower, both in dividing up whole beefs between consumers and in communicating the cutting instructions to the processor.

Even if they offer set packages, producers will usually modify these selected cuts within reason. It doesn't hurt to ask. Also, many consumers don't take their share of the offal, so if you're a big liver 'n' onions fan or want the tongue or tail, you should order it. If you're the only one to do so, you may be in for a bonus.

The following is an example of the cuts included in a "quarter-beef package" offered by a local grower:

Steaks:

4 tenderloin steaks   4 sirloin steaks
6 New York strip steaks   4 sirloin tip steaks
6 rib-eye steaks   6 cube steaks

Roasts:

4 chuck roasts (about 3 pounds each)
2 rump roasts
4 top round

Ground Beef:

Approximately 36–40 pounds of lean ground beef in
    1-pound packages

Specialty Cuts:

> A combination of brisket(s), tri-tip(s), and/or flank(s) cut to
> approximately 3 pounds each, two pieces total
> 2 pounds of stir-fry/fajita meat
> 2 pounds of kabob meat
> 4–6 pounds of short ribs
> 2 pounds of soup bones

Even if you aren't given much leeway, how might you customize the order above to give you just the meat you want? Well, say I don't want the temptation of making chicken fried steak (with cream gravy and biscuits!), so I want to avoid ordering cube steaks. I'll end up with more meat from the round. More kabob meat? Might I trade my sirloin tip steaks for a roast? Can I have bone-in roasts? I reckon the short ribs will come English-style, cut with the bone, not against it in thin slices; I wonder if I could have half of them flanken-style? Can I have the tongue? Oxtail? Is there beef shin for the taking? Can some of my meat be made into pepperoni—can you make that spicy pepperoni, please?

You see where I'm going with this. Now you have just enough information to make you a dangerous, I mean an informed, consumer. Knowing the primals and being able to envision where the cuts hail from doesn't just help you when ordering a whole animal, either. It will help guide your purchases and cooking methods for all meat.

How Much Will My Beef Cost and How Much Will I Get?

I am not going to lie to you: Buying a whole beef is a very pricey proposition, in the thousands of dollars. For a local grass-fed and -finished beef, farmers charge about the same amount that I paid for my 1965 Falcon in good condition. Whether you find a classic car or a freezer full of meat a more alluring deal I can't guess, though if you do your homework the beef should give you a much more reliable return on investment. The beef will also take up less room than the car, though you will require a committed chest or standing freezer to hold a haul this big, so make sure you factor that into the cost. The final cost of your beef will depend on many factors, including the animal's diet and any formal certification, your area of the country, and the processing fees for butchering and cutting and wrapping the meat.

Because beef carcasses are so large, however, it is possible to buy whole, half-, quarter-, and eighth-shares directly from the growers, or you can go in with friends and divide the beef up yourself instead of depending on the producer to do it. Because there are so many options for carving up a beef, below I outline more precisely what to expect for each portion—whole, half, quarter, eighth—of a beef you buy. But no matter what portion of a whole animal you buy, you will need to put up a substantial deposit when you place your order.

In general, expect to pay around $3.50–$5.50 per pound for a beef when ordering the whole animal. I've found some for as low as $2.40 in my area and some nearing $6 (that's a precious cow, my friend). Let's run through two examples, each scenario based on actual fee structures from small-scale producers, one here on the West Coast and one in the Midwest.

Producer Good Cow charges $3.89 per pound for a half-beef, including the kill fee and cut-and-wrap. On their website, the average hanging weight of a half-beef from Good Cow is listed as 300 pounds.

(average hanging weight) 300 x
(inclusive price per pound) $3.89 = $1,167 total cost

Now, over at Fat Cow Farm, they charge $2.40 per pound for a whole beef and slightly more for a half, but that price is still just $2.45 based on hanging weight. What a deal! But wait: Processing and slaughter are not included. Unless you specify a different processor, they work with a USDA processor, Quality Meats. Quality Meats charges $.45 per pound for processing, with an additional $60 disposal and kill charge per order. Upon inquiry, you are told that hanging weights at Fat Cow Farm range from 300 to 380 pounds for a half-beef.

You have to calculate two costs here and total them together. The first is from the grower for the meat. Let's start with an average weight of 340 pounds for a half-beef. We'll use the formula we used before to calculate our costs from the grower:

$$340 \times \$2.45 = \$833$$

We also need to pay Quality Meats to slaughter the animal, age the meat, cut and wrap it, and freeze it:

$$(340 \times \$0.45) + \$60 = \$213$$
$$\$833 + \$213 = \$1{,}046 \text{ total cost}$$

You will note that in the end the two costs are not as far apart as they initially seemed they might be. Make sure you read all the terms carefully, or ask all the right questions, to ensure you aren't surprised by hidden costs.

Though I have listed the most common pricing methods that smaller producers use, one grower I found does the math for you:

Whole Beef $2,000 (385–400 pounds)
Half-Beef $1,050 (185–200 pounds)
Quarter-Beef $575 (85–100 pounds)

At first I was flummoxed by these tiny, expensive cattle! Then I read the fine print. They charge a set fee for a whole, half-, or quarter-beef; the weight range given is the amount of meat you will take home.

Questions to Ask Your Grower:

» Is the cost calculated by live weight or hanging weight, or is there a set cost per whole, half, or quarter?
» How much deposit is required?
» What breed are your cattle? Why do you raise that breed?
» What is the average hanging weight of your animals?
» Is the slaughter and processing included in the fee, or will I pay the processor separately?
» Will the meat be dry-aged?

How Much Meat Will I Get?

It's a cow, after all, so you know it will be substantial. But from a 1,200-pound beef on the hoof, let's assume you'd end up with a hanging weight of 732 pounds (61 percent of live weight—not a bad average to go on according to a couple of university publications).

For this average whole beef carcass, cut into boneless steaks and roasts, closely trimmed of all fat, with lean ground beef, you'd take home 62 percent of the hanging weight for a total end weight of meat of 454 pounds.

If you choose bone-in steaks and roasts, with average fat trim, and regular ground beef, you raise that to 71 percent of the hanging weight, for a finished cut weight of 520 pounds.

You'll probably end up somewhere in the middle. You want the soup bones, for example, but won't take the liver. You want your tenderloin whole and tied, but leave the bones in your roasts. Let's say that, for this average whole beef, cut into some boneless and some bone-in steaks and roasts, closely trimmed of all fat, with regular ground beef, you'd take home 67 percent of that carcass for a total end weight of meat: 490 pounds.

Only interested in a half a beef? (You're getting smarter.) Because of all the varying factors, let's assume your cut-out weight or take-home meat will vary from 62 to 71 percent of the hanging weight. If we apply this percentage to the average hanging weights we used in calculating cost, following is what we end up with for a half-beef with hanging weight between 300 and 340 pounds. At the low end:

300 pounds hanging weight x 62–71 percent =
186–213 pounds of take-home beef

And at the high end:

340 pounds hanging weight x 62–71 percent =
211–241 pounds of take-home beef

These numbers, you will note, are fairly close to the spreads given by the grower who priced his beefs by take-home weight. However, if I've learned anything, it's that cattle come in more sizes than pants. Asking your individual producer is the fastest, easiest way to get the best estimate based on their breed and average slaughter weight. Now you have the tools, however, to do some good guesstimation while you look around at producers. You can also do some pretty fancy talking at the farmers market. Listen to you, talking hanging weight and carcass cutting yield—Uncle Dave would be proud.

How Much Freezer Space Do I Need?

A whole beef is good for up to 500 pounds of take-home yield. You will need either an entire chest freezer plus a portion of a standing freezer, or collaboration with friends.

A half-beef yields approximately 225 pounds and requires the equivalent of a small chest freezer or standing freezer.

A quarter-beef yields approximately 110 pounds. Some producers claim a quarter-beef will fill the freezer space of a standard refrigerator/freezer unit, but I will tell you that Uncle Dave's cow-quarter filled two-thirds of a 13.7 cubic-foot upright freezer, so I say they're lying (Cousin Kate says they're selling you an old, skinny cow).

An eighth-beef gives you approximately 50 pounds of take-home yield and requires approximately 2.25 cubic feet of freezer space, or the equivalent of the freezer space in a smaller refrigerator/freezer unit. For an apartment fridge this means the entire thing. Like no room for ice. How will you make your martinis?

Remember that how you have your beef cut will affect how much space you need. Steaks are flat; they stack. Likewise with ground beef. Roasts take up more room.

Who Will Kill My Cow and How?

If your cow is not going to end up as kosher or halal beef, it will most likely be killed by being shot in the head or stunned with a captive-bolt stunner before being cut. The USDA's Food Safety and Inspection Service (FSIS) guidelines call for rendering the cow insensible with a single blow prior to hoisting and slaughter (ritual slaughter is exempt). The FSIS also calls for many of the other aspects of humane slaughter we have already covered, including minimizing discomfort in transport, giving cattle continuous access to water, keeping them calm, and protecting them from extreme heat and cold. Remember, not only are rough handling and sloppy slaughter methods inhumane, but they result in lower-quality meat. Roughly handled cattle may translate into bruised meat, while stressed animals can become what are known as "dark-cutters," or carcasses marked by dark meat that is both tougher and more prone to spoilage. (For more information on the chemistry behind that reaction, see "Who Will Kill My Pig and How?" in Chapter 7, "Pig.").

When Will It Be Ready?

Larger producers may be able to fill a smaller beef order on a relatively tight timeline, by which I mean about a month from order to delivered meat. This includes aging time, remember. More likely, especially if you're working with a smaller producer, you may need to reserve your beef months in advance for a specific slaughter date. When you're working with whole animals, the glory doesn't come in the immediate gratification. It comes later, on that rainy, windy day when, instead of going to the store, you go home and eat that wonderful meat from your very special cow you put in to braise that morning, or pull out hamburger for a quick defrost. Just like changing how we grow and eat meat, the process is long, but worth the journey.

A Note on Dry-Aged Beef

Dry-aging is a process by which butchered portions of a beef carcass are hung to rest in a humid, cool environment. During the days or weeks that beef ages, it loses moisture and weight. Enzymes begin their work to break down the meat, producing amino acids, fatty acids, and sugars, and concentrating the flavor. Generally speaking, most butchers will dry-age a carcass for somewhere between seven and fourteen days, which is usually enough to increase the flavor profile and improve tenderness. Because of the time spent aging the meat and the loss of weight, dry-aged beef is more expensive to produce.

If you plan to buy grass-fed or other leaner meat with less fat cover, ask if your processor ages for more than a week, as longer aging times for lean beef can adversely affect the end product. If your processor doesn't or won't age your beef before cutting and wrapping it, you can re-create some of the benefits by allowing individual cuts to rest unwrapped on a rack in your refrigerator for a day or so before cooking. I find less difference with "wet aging," which is aging meat at a cool temperature sealed in plastic, as is done with supermarket meat.

RECIPES FOR COW

French Onion Soup

There is a lot to love about a gorgeous, molten bowl of French onion soup—the deeply caramelized onions nestled in a rich thyme-scented broth, and of course the cheese, the favorite part for kids and adults alike (though some adults won't admit it). Why would you order a soup this good in a bistro but never endeavor to make it at home? Maybe because great beef broth is the foundation of the recipe, and while lots of folks now freeze their chicken carcasses and wing tips for stock, far fewer have soup bones lying around. But now you do! (Note: The soup bones from Uncle Dave's cow were extraordinarily meaty. As my brother Jeff said, "They're more like beef shanks than soup bones!" You can use a bit more if your cow has skimpy legs to ensure you get an unctuous broth.)

SERVES 6–8

For the broth:

    Olive oil for browning

    4 pounds beef soup bones

    1 celery heart, rinsed

    1 large carrot, scraped and halved

    2 red onions, quartered

    2 shallots, halved

    1 tablespoon black peppercorns

    2 sprigs fresh thyme

In a large stockpot or Dutch oven, heat a couple of tablespoons or so of olive oil over a high flame. Add the soup bones, and knock them around a little until they are brown and start to caramelize, about 15–20 minutes.

Lower the heat to medium-high. Deglaze with a splash of water (or sherry), and loosen up any browned bits on the bottom of the pan. Add the remaining ingredients and cover with cold water. Bring back to a boil, and then lower the heat to a slow simmer. The bones will throw off scum at first; skim this off with a mesh strainer. As the broth cooks, check back every so often and strain again as needed.

Cook for at least 3–4 hours, more if you can do it. Add more water as needed. I like to cook it all day Saturday, strain it, pressing lightly on the solids, and then refrigerate it overnight (put the veggies in the compost and give the bones to the dogs). When Sunday rolls around, you're less than an hour away from soup. Technically, you should be able to skim the fat after the broth cools, but my broth chills into more of an aspic because of all the gelatin in the bones. It's gorgeous.

> For the soup:
> Olive oil for frying
> 6 yellow onions, thinly sliced
> 3 sprigs fresh thyme
> ½ cup dry sherry
> 4 cups hot beef broth
> ½ baguette, sliced crosswise in ½-inch slices
> 6 ounces Gruyère cheese, grated

Preheat the oven to 350°F.

In a large stockpot, heat a couple of tablespoons of olive oil over a medium-high flame. Add the onions and toss them to coat with the oil. Season with salt and pepper, and add the thyme sprigs. Cook, stirring frequently, until the onions begin to brown and turn limp.

Add the sherry, and stir to bring up any browned bits. Cook the sherry down, then reduce the heat to low. Cover and cook for 30–40 minutes more, or until the onions are deeply golden, sticky, and caramelized. Remove the thyme stems.

Add the hot broth, and stir. Check the seasoning. Adjust the heat so the soup stays at a low simmer, and cook for 15 minutes to marry the flavors.

Place the baguette slices on a sheet pan. Toast the bread in the oven, flipping once, until it is just crisp. Remove the bread and turn the oven to broil.

Take out those beautiful hand-me-down French onion soup bowls of your mother's that you never use, preferably ones with handles. Line a sheet pan with parchment (you'll thank me for this later), and place the bowls on the sheet pan. Ladle about a cup of soup into each one, making sure to include plenty of onions in each. Top each with a slice of toasted bread, and sprinkle lavishly with the grated Gruyère.

Run the soup under the broiler until the cheese is golden brown and bubbling over the sides. Serve at once. If serving to children, park their bowls for a few minutes to allow them to cool (extra slices of bread with broiled Gruyère make a nice snack while they're waiting).

Stacked Beef Enchiladas

These are enchiladas in the New Mexican style, easier than and just as delicious as frying and saucing and rolling, as other recipes ask you to do. And prunes, really? Really. The fruit and cocoa pair so nicely with the chiles that you'll rethink prunes forevermore.

To make the sauce, you must use a food mill to process the braising liquid; trying to push it through a sieve is both futile and silly. (Note: For convenience, feel free to cook the meat all day in a slow cooker rather than in the oven.)

SERVES 4 GENEROUSLY

Olive oil for searing

3–4–pound sirloin tip roast, or other cut from the chuck or round

1 small yellow onion, chopped

6 garlic cloves, sliced

28-ounce can San Marzano tomatoes

1 cinnamon stick

1 tablespoon hot ancho chile powder or chipotle chile powder

12 guajillo chiles, toasted

Pinch Mexican oregano

2 teaspoons unsweetened cocoa powder, or a small chunk of unsweetened chocolate, grated

8 prunes

⅓ cup raisins

Lard or vegetable oil for frying
16 corn tortillas
2 cups crumbled queso fresco, or shredded aged Monterey Jack

Preheat the oven to 300°F.

Film the bottom of a Dutch oven with olive oil, and heat over a high flame. Salt and pepper the roast liberally, then sear it on both sides. Remove the roast and set it aside.

Add the onion and sauté until soft. Add the garlic, and toss to coat with the oil. Add the tomatoes, cinnamon, chile powder, chiles, oregano, cocoa powder, prunes, and raisins. Stir to combine, and bring to a simmer. Add the roast and enough water to cover the meat about halfway.

Cover the pot and place in the oven. Braise for about 5 hours, turning the meat once or twice if you're so inclined. Remove the meat and allow it to cool slightly, then shred it with your hands or two forks. (Note: All the steps up to this point can be done a day ahead of time.)

Preheat the oven to 450°F.

Run the braising liquid through a food mill and into a clean saucepan. Reduce the sauce over medium-high heat to thicken it slightly.

Heat ½ inch of lard or oil in a wide saucepan or Dutch oven over a high flame, until a chopstick or the handle of a wooden spoon bubbles on contact. Pass the tortillas, one at a time, through the hot oil just until they blister and puff. Remove them at once and drain them thoroughly.

Place 4 tortillas on a cookie sheet or half-sheet pan. Top each with a bit of meat, a drizzle of sauce, and a bit of cheese, then top each with another tortilla. Repeat until you have 4 stacks with 3 layers of meat and cheese and 4 tortillas in each. Top each stack with a generous drizzle of sauce and a shower of cheese. Bake until heated through and the cheese is melted and bubbly, about 10 minutes.

Ethiopian Steak Tartare (*Kitfo*)

One of the glories of having meat you totally trust, especially beef imbued with that deep, grassy flavor, is that you feel safe eating it raw. Ethiopian beef tartare, called kitfo, is a complete pleasure—intriguingly spiced, rich, and satiny. I know what you're going to think—one cup butter? You could get away with less, but with less concomitant enjoyment.

I adore Ethiopian food, both the tastes and the presentation. The rich stews and vegetables are dolloped on a large circular platter lined with an enormous spongy, fermented pancake called injera. You tear off pieces of extra injera and use them to pick up and combine tastes and bits and convey them to your mouth. When you're done, you eat the injera that lines the platter, now soaked with deliciousness. In my efforts to make Ethiopian food at home, I've had to adapt to the American kitchen (and to our Northwest climate, which lacks the radiant Ethiopian sun for drying out spice pastes for storage). Below is the pleasing result of much trial and error on my part, suitable for making in areas where sunshine is often in short supply. As for the injera, my advice is, buy it. If you can't, substitute something altogether different, like a baguette, naan, or chapatis.

SERVES 6

½ teaspoon fenugreek seeds

½ cup ground New Mexican chiles

¼ cup paprika (sweet-hot if possible)

1 tablespoon salt

1 teaspoon ground ginger

1 teaspoon onion powder

Seeds from 3 green cardamom pods

1 teaspoon coriander seed

¼ teaspoon garlic powder

4 passes of a nutmeg on a microplane grater

⅛ teaspoon each ground clove, cinnamon, and allspice

2 teaspoons cayenne, or 1 tablespoon Aleppo pepper

1 cup butter

2 cups 2% cottage cheese

2 pounds ground beef, or tenderloin, cut into small dice

For the cooked variation:
    2 cups chopped yellow onion
    A couple of splashes of red wine

Combine all of the spices in a coffee grinder, and grind until finely powdered. Melt the butter in a saucepan over medium-low heat, and cook until it foams. Skim off the foam and add the spices, stirring until the butter is aromatic. Reserve 3 tablespoons of the spiced butter and add to the cottage cheese, along with a small pinch of salt to taste. Stir to combine.

Remove the saucepan from the heat and add the meat, stirring to combine and coat the meat with the butter and spices. Immediately turn the mixture out onto an injera-lined plate or platter. Spoon the spiced cheese next to the meat, and serve immediately with extra injera or bread.

High Squeamishness Index Variation: Once when Husband Hendel and I ordered kitfo in a Seattle Ethiopian restaurant, the server refused to bring it to us raw. They only had five things on the menu, so I would have been surprised if it wasn't ordered with frequency. However, she insisted we really wanted it cooked at least a little, like most (non-Ethiopian) customers prefer. But if you're one of those steak-done-medium-well people and I can't convince you otherwise, then I think you should change things up completely. Brown the meat in a little butter and set aside. Cook 2 cups of chopped onion in a dry pan (it's the Ethiopian way) until it's limp and nearly browned. Add the butter and spices and cook, stirring, until the butter melts. Add a couple of splashes of red wine, then the beef and a bit of water. Cover and cook until the beef is tender. It's not kitfo, but it's good.

Pastrami

Pastrami doesn't fall, sliced, from the heavens. It starts out as brisket or top round, and if you ordered part of a beef I know you've got some of that in your freezer. I use top round.

Pastrami making is an adventure; it's delicious, but it's not something you do on a whim. It needs to be cured (or "corned"), then smoked at a low temperature, then cooked. (Some recipes smoke the meat until it is fully cooked, but because that can take up to ten hours, I prefer this combined method that I adapted from a recipe in Cured *by Lindy Wildsmith.) And when you're done, unless you have a really, really sharp knife you might also get the urge to buy a commercial slicer, and if you do that, you can make tremendously expensive yet delicious Reuben sandwiches to your heart's content. Incidentally, this recipe uses no curing salt (also called* pink salt *or* Prague powder), *though you could sub some for some of the sea salt if you want your meat to stay pink.*

SERVES 6–8

For the cure:
  2 bay leaves
  6 garlic cloves
  Pinch red chile flakes
  1 tablespoon each black peppercorns, white peppercorns, coriander
      seed, mustard seed, paprika, and juniper berries
  4 pods green cardamom, cracked and seeds removed
  1-inch piece fresh ginger, peeled and sliced
  5 tablespoons coarse sea salt
  3 tablespoons turbinado sugar
  5 pounds grass-fed brisket, trimmed, or top round roast

Combine all of the seasonings in a spice grinder, food processor, or blender, and blend to a rough paste. (I used a mortar and pestle and regretted it later. Whatever state you reduce this blend to will form the spice crust on the outside of the pastrami, and big pieces of bay leaf are a little too, let's say, toothsome.) Rub the meat all over with this mixture and place in a resealable bag. Squish out all the air and seal. Place the bag on a tray or platter, cover it with another plate, weight it with a couple of foil-wrapped bricks or two 28-ounce cans of tomatoes, and refrigerate.

Once or twice a day, flip the meat over and squish the bag around a little. Do this for at least 10 and as many as 21 days.

The day before you want to finish the pastrami, remove the meat from the bag, and place it on a rack set over a jelly roll pan or plate. Allow to air-dry in the fridge overnight.

Cold-smoke the pastrami (keep temperature around or below 90°F) for about 1 hour—just enough to get a hint of smoke flavor. Fill a large pot with water and bring to a simmer. Place the pastrami in a plastic roasting bag and remove as much air as possible. Place the bag in the water bath, and simmer for about 90 minutes, or until the meat's internal temperature reaches 165°F. Remove the bag from the water bath, wrap the meat in foil, and chill it completely.

Barnyard Hand Pies

I really wanted to call these Cow Pies, but I realized that I made these turnovers using lard I'd rendered from my pig, then stuffed them full of Uncle Dave's cow and someone else's chickens' eggs, so the whole farm was involved. The proper name for these little morsels is really empanadas, which is where the fairly classic beef-olive-egg combo comes from. But you don't have to stuff them with beef, or even animal at all. Potato, onion, and peppers are nice, too, as are shrimp or tuna and olive. Your own Spanish-Style Chorizo (see the Pig recipes section) would be delicious in place of beef, in which case I would leave out the eggs. The possibilities are endless! The lard-butter crust from A Summer Pie (in Chapter 7, "Pig") may be substituted for this pastry dough if you like.

MAKES 12

> For the pastry:
>> 2 cups flour
>> 1 teaspoon kosher salt
>> ½ cup cold lard (or butter)
>> ¼ cup dry white wine
>> 2 eggs

For the filling:

- 1 tablespoon lard or olive oil
- 1 pound ground beef
- 2 green bell peppers, seeded and diced
- 2 cloves garlic, peeled and chopped
- 1 large yellow onion, minced
- 2 large Yukon gold potatoes, peeled and cut into ¼-inch dice
- 2 teaspoons smoked paprika
- 3 sprigs fresh oregano, leaves stripped and chopped
- ½ cup tomato sauce
- ½ cup green olives, pitted and roughly chopped
- 2 hard-cooked eggs, roughly chopped

To make the dough:

Pulse the flour and salt together in a food processor. Add the lard or butter, and pulse a few times, or until the lard is just distributed. Blend together the wine and eggs and pour into the food processor. Pulse just until the dough comes together. Dump the mixture onto a board, and knead just until it forms a ball. Separate the dough into 2 disks, and wrap them well in plastic. Chill them in the refrigerator for at least 1 hour, or up to overnight.

To make the filling:

Heat the lard or oil in a sauté pan, and add the ground beef and a pinch of salt. Cook over medium-high heat, stirring, until the beef is browned. Remove the beef, reserving the fat in the pan. Add the green peppers, garlic, onion, and a pinch of salt to the pan (this mixture is called a *sofrito*), and cook for 5 to 7 minutes, or until the vegetables are soft. Add the potatoes, cover the pan, and cook, stirring occasionally, until the potatoes are tender. Add the beef, paprika, oregano, and tomato sauce, and cook until slightly thickened, about 5 minutes. Remove the mixture to a bowl, and stir in the olives and egg.

To make the pies:

Preheat the oven to 400°F.

Roll out the dough to ¼-inch thickness, and cut out 4-inch circles with a large biscuit cutter. Place 2 tablespoons of filling on each, wet the edges, fold the dough over, and press to seal in a half-moon shape. Place the pies on sheet pans, and bake them for about 20 minutes, or until golden. (Alternatively, halve the size and deep fry them in lard if you dare. Heaven.)

Smoked Pot Roast

If you buy a cow, you will become acquainted with pot roast. And that's not a bad thing. Pot roast is incredibly versatile. Braise with wine, braise with water, add chunky veggies, go the picadillo route—you've got options. But chances are, once you're a ways into that cow you might want more options than you initially thought. For variety, let me offer this recipe for smoked pot roast from Bette Taylor, an old family friend. The recipe works beautifully with pork, too, though I've adjusted the spices and quantities to make it a little less sweet for the beef. Increase the sugar if you want to use the rub with pork.

SERVES 4–6

¾ cup (packed) brown sugar

½ cup each kosher salt, garlic powder, and sweet paprika

3 tablespoons onion powder

1 tablespoon dry Creole seasoning

1 tablespoon cayenne (or more to taste)

3 tablespoons chili powder

1 tablespoon black pepper

3 tablespoons Dijon mustard

8–9 pounds arm steak, round steak, or brisket

Combine all ingredients but the mustard and the meat. Rub the meat all over with the mustard, then smear it well with the spice mixture. Wrap the meat in plastic, and allow it to sit overnight, refrigerated.

Preheat the smoker to 220–250°F.

Remove the meat from the refrigerator 1 hour before cooking, and let it sit uncovered on the counter to take off the chill. Smoke it for 4 hours.

Preheat the oven to 250°F.

Wrap the meat well in foil, and cook it for another 4 hours in the oven. Let it rest for 30 minutes before serving.

Sweet-Hot Jerky

Brother Dee graciously offered his incredible jerky recipe—for the first time ever— after much cajoling on the part of his sister. Dee usually uses venison or duck for this, but, as he says, "It's been proven on many meats," and my attempts with beef were equally delicious if not quite as exotic. He makes his spicy enough that "fifty percent of people can't handle it, which is kind of the point." Otherwise, kids will eat it faster than you can make it. Dee cures his jerky in his jerry-rigged backyard refrigerator smoker. If you have a smaller smoker and need to make the jerky in batches, you can smoke for taste and then finish drying it in the oven or in a food dehydrator. Make sure you maintain the low temperature: You want the jerky to cure, not cook. Cooking is unnecessary, and it will ruin the texture.

MAKES APPROXIMATELY 8 POUNDS

20 pounds grass-fed beef such as round or chuck, or any other lean meat such as venison

6 tablespoons salt

6 teaspoons Prague Powder #1 or Morton's TenderQuick

1¼ cups (packed) brown sugar

1¼ cups white sugar

1 cup corn syrup

8 teaspoons each freshly ground black pepper, cayenne pepper, and crushed red pepper flakes and seeds

4 tablespoons garlic powder

2 tablespoons powdered ginger

4 cups soy sauce

4 cups pineapple juice

Place the meat in the freezer (or thaw it) until very firm but pliable. Slice ¼–⅜ inch thick while still partially frozen, removing every trace of fat and sinew.

Mix all the other ingredients well in a large nonreactive bowl. Add the sliced meat, and stir until it is evenly coated. Marinate in the refrigerator for at least 24 and up to 48 hours, stirring 3 or 4 times a day.

Drain the meat, and lay it out on the smoker racks. Dry it in the smoker at 90–100°F, not allowing the smoker to get any hotter. Smoke it for 18–36 hours, depending on humidity and the thickness of the meat, until the meat is the consistency of stiff shoe leather. The usual formula uses 2 to 4 pans of wood chips, depending on the desired taste, though the jerky can be smoked and then finished in the oven or a food dehydrator. Store in an airtight container.

Those Pretty Pink Salts Aren't for Bathing

Curing salts, also known as tinted cures, Insta-cures, or pink curing salts, are used to inhibit the growth of bad bacteria—such as *Clostridium botulinum*, which causes botulism—and to help preserve cured meats. They also keep meat products a lovely rosy hue. Prague powder #1 is used in cured meats such as hot dogs or pastrami, while Prague powder #2 is used for slow- or dry-cured hard sausages and meats. It's essential you use neither too much nor too little in recipes, and this is the very reason the salts are colored a pretty pink—so that you don't mistake them for regular table salt. General guidelines call for one level teaspoon of cure for every five pounds of ground meat, though the instructions in individual recipes vary and should be followed carefully. Better yet, grab a copy of sausage king Rytek Kutas's book *Great Sausage Recipes and Meat Curing*, or another of the charcuterie-minded tomes listed in the back of this book, and read them through and through before starting your own cured meat adventures.

Braised Beef and Daikon

If you don't specifically request beef shin or shank, and your butcher is not Chinese, the meat will probably be ground and the bones given to you for soup. But beef shin is a delicious cut for long braises. The heavily used leg muscle has loads of connective tissue that translates into a melty mouthfeel in the finished product. This is a classic Chinese dish in which the large, long, white Asian radish (daikon in Japanese) plays a starring role, becoming tender and suave during its time in the pot. Children will be initially suspicious, but don't call it "radish" and they might give it a go. If you don't want to shallow-fry the meat before braising, just brown it well in a little oil. Use a heavy pot that will hold the heat, and do it in batches to make sure you sear the meat instead of steaming it.

SERVES 4–6 AS PART OF A CHINESE MENU

1½ pounds beef shin or shank

2 tablespoons Shaoxing wine or dry sherry

1 tablespoon cornstarch

6 tablespoons soy sauce

Oil for frying (preferably vegetable or peanut)

1 leek

1 large daikon, about 2 pounds

2 garlic cloves, peeled and smashed

2 whole star anise

5 thick slices fresh ginger

2 dried red chiles (optional)

Cut the meat into 1½-inch pieces and place in a bowl. Sprinkle with the Shaoxing wine, the cornstarch, and 1 tablespoon of the soy sauce. Toss to coat, and set aside.

Heat 2 inches of oil in a Dutch oven or heavy pot, until vigorous bubbles form around a wooden chopstick inserted into it. Add the meat and fry until well browned. Remove the meat and drain. Pour off all but 2 tablespoons of the oil.

Thinly slice the leek, white part only, on a diagonal. Rinse the slices well and dry them thoroughly. Trim and peel the daikon, then cut crosswise into rough 2-inch chunks. Halve the largest pieces (the ones closest to the stem), and round off the edges of each with a paring knife.

Heat the pot over a medium-high flame. Add the leek, garlic, anise, ginger, and chiles (if using). Cook, stirring constantly, for a few minutes to release the flavors. Add the meat, 2 tablespoons of soy sauce, and 6 cups of water. Bring to a boil, then lower to a simmer and cover. Cook, skimming occasionally, for 2–3 hours, or until the beef is tender.

Uncover the pot and add the daikon. Simmer until the liquid has reduced roughly by half, then taste for seasoning and add the remaining soy sauce if you like. Simmer gently until the daikon is very tender.

Guncle Dee's Super-Secret Famous Ribs

These aren't a secret among the Millers, of course. I've had the good fortune to devour them on any number of occasions, but Dee wanted me to tell you that he has never before given anyone the recipe. Once you try them, you'll feel appropriately grateful. Dee grills, smokes, and then grills them again, but he's like that. If you don't have a smoker, you can bake them longer and grill them for a little smoky hit, but you'll risk them falling apart on the grill. If you like them saucy, paint the ribs with your favorite sauce before giving them a final turn on the grill (Dee prefers McCade's, from Oklahoma). You can make these with pork ribs, but my dad always liked beef ribs best, so this one's for him.

SERVES 4–6

½ cup mild Chimayo chile powder
½ cup hot Chimayo chile powder
1 cup sweet paprika
¼ cup each freshly ground white pepper, freshly ground black pepper,
    onion powder, garlic powder, kosher salt, and granulated sugar
1 rack beef ribs, trimmed

Combine the rub ingredients. This amount is enough for at least two rib adventures; store in an airtight container.

The night before cooking, peel off the "fell," the opaque membrane on the back side of the rack. Massage the ribs well with a generous amount of rub. Cover the meat and refrigerate overnight.

Heat a grill to high, and get your smoker ready. Sear the rack on the grill until the meat is slightly crisped, taking care not to burn the chile powder, or it will turn bitter.

Remove the rack from the grill, and place it in the smoker. Smoke at about 200°F, for about 4 hours. Your goal is to infuse the meat with flavor and cook it until tender.

Glaze the meat with your favorite sauce, and run it over the grill just until the sauce is set.

If you don't have a smoker, wrap the rack in foil after the initial grilling. Bake it in a 200°–250°F oven for 3 hours, or until the meat is falling off the bone. Serve as is, or apply your favorite sauce and run it quickly under the broiler to caramelize the sauce.

Rainy Sunday Chili

It was January. Husband Hendel said he wanted noodles in broth or fish for dinner, Asian style. But he was going on an epic bike ride, and I knew he'd be starving when he got back. I was thinking chili—I'd already soaked a pound of black beans overnight, something I rarely have the foresight to do. As it turned out, I made the right choice.

But, trying to please everyone, I made the chili too spicy for the kids and not spicy enough for me. (To make up for it, I made them pie.) Adjust accordingly.

SERVES 4–6

6 guajillo chiles

6 chiles de arbol

2 tablespoons cocoa powder

1 28-ounce can San Marzano tomatoes

1½ pounds beef stew meat, cut into ribbons

Olive oil for frying

1 medium yellow onion, minced

5 garlic cloves, minced

1½ teaspoons cumin seeds, toasted and ground

1 teaspoon dried Mexican oregano

2 tablespoons pure ancho chile powder

1 pound black beans, soaked overnight and drained

Combine the chiles and 2 cups of water in a small saucepan, and bring to a boil. Reduce the heat, and simmer for 10 minutes. Remove the saucepan from the heat, cover, and allow the chiles to soften for 10 minutes.

Place the chiles in a blender with enough soaking liquid to loosen the blade, and purée them. Add the cocoa powder and tomatoes, and blend until smooth.

Season the beef with salt and freshly ground pepper. Heat a few tablespoons of olive oil in a Dutch oven or heavy pot over a medium-high flame, and brown the meat. Add the onion and cook, stirring, frequently, until limp. Add the garlic, and cook for another minute. Add the cumin, oregano, and chile powder, and stir to combine. Add the beans and the tomato–chile purée.

Add enough water to cover the beans by 1 inch. Cover the pot, and simmer for 2–3 hours, or until the beans and meat are tender. Taste for salt and serve. (The flavor improves over the passing days.)

Jeff's Swedish Meatball Army

When Jeff told me that three pounds of meat went into these, I balked a little. That's a lot of meatballs, even for a party to honor the Saint of Lutefisk or whatever it is my people might celebrate. But Swedish meatballs are delicious and, as Jeff says, if you're going to make meatballs, you might as well make meatballs. Eat them plain, or make a roux and whisk in some heavy cream, add fresh dill and maybe sautéed mushrooms. My mother serves them over noodles. Cousin Kate says they freeze like a dream.

SERVES AN ARMY

1 pound each ground beef, ground pork, and country pork sausage

1 cup mashed potato flakes

1 cup breadcrumbs

2 eggs, lightly beaten

1 tablespoon fine sea salt

1 tablespoon brown sugar

½ teaspoon each freshly ground black pepper, freshly ground nutmeg, ground clove, and ground allspice

2 teaspoons dried dill

Oil for browning

Preheat the oven to 400°F.

Gently combine all the ingredients, and shape into compact balls. Brown them in a little oil over medium-high heat in batches, then bake for 15 minutes.

Jeff's Swedish Meatball Army

Thai Minced Beef Salad

One of my favorite Thai dishes is larb gai, *a barely cooked, herb-packed chicken dish that's both flavorful and light, all at once. The loose salad is scooped up with lettuce leaves or cabbage, or served with sticky rice. Though it's delicious with chicken, the flavor and texture of grass-fed beef work beautifully as well; you could even use pork. This may be a bit herbaceous for kids, and too spicy for sure. The toasted rice powder takes just a minute to make and is essential to the dish, acting as both a binder and a flavor agent.*

SERVES 4 AS PART OF AN ASIAN-STYLE MEAL WITH RICE

2 tablespoons raw sticky rice

4 black peppercorns

1 pound lean ground grass-fed beef

1 teaspoon brown sugar

Juice of 2 limes

2 tablespoons fish sauce

3 fresh bird's-eye chiles, minced

3 large shallots, peeled and finely chopped

2 green onions, both green and white parts

¼ cup mint, coarsely chopped, plus additional sprigs for garnish

½ bunch cilantro leaves, roughly chopped, plus additional sprigs for garnish

3 sprigs holy basil

½ head cabbage or lettuce, if desired, cored for serving

Heat the rice in a dry cast-iron skillet over a medium flame. Roast it, shaking the skillet, until deeply golden and fragrant. Remove the pan from the heat, and pour the rice into a mortar. Add the peppercorns, and grind the mixture to the texture of course sand.

Bring a generous ½ cup of water to a boil in a wok. Add the meat and cook, stirring to break up any lumps, until it is still a little pink. Add the rice powder and stir well.

Add the sugar, lime juice, fish sauce, and chiles, and toss to coat. At the last second, add the shallots, green onions, and chopped herbs. Turn out the mixture into a shallow bowl, and garnish with herb sprigs. Serve with cabbage and sticky rice if desired.

Faux *Pho*

I have the good fortune to live in a part of Seattle that is literally cheek by jowl with pho houses, and thankfully not the kind that tame their Vietnamese soups for white people. I'm talking about bowls full of special chewy noodles, twin noodles, beef tendon, and braised duck with house-pickled young guava. Bowls filled with pig's blood and hearts, and one whose secret I have yet to crack: a bright orange "semi-fluid" soup that begins its time with you as nearly a gel and liquefies as time goes on. I'm not sure I want to know what produces that effect, but I admire the trick. With this wealth mere blocks away in every direction, it almost seemed silly to make Vietnamese-style soup at home. Except that Uncle Dave's cow lived in the freezer, and we didn't want to go out. Romeo protested my calling this soup pho because I am not Vietnamese, but the boys slurped it up and we all ate our fill. And thus faux pho was born.

SERVES 6

3 large shallots, 2 cut in half lengthwise and 1 peeled and thinly sliced

2 medium onions, cut in half lengthwise

1 knob of fresh ginger, cut in 3 pieces on the bias

About 1½ pounds beef soup bones

2 large carrots, halved

Small handful black peppercorns (about 20)

3–4 star anise

3 cinnamon sticks

7 cloves

You're not Vietnamese.

2 tablespoons whole coriander seeds, cracked

Lump of rock sugar, or 1 tablespoon brown sugar

Oil or lard for browning

1 pound beef stew meat, from the chuck

2 tablespoons fish sauce

2 packages *banh pho* or rice sticks

1 pound flank steak, slightly frozen and sliced thinly against the grain

Handful of basil sprigs

Handful of cilantro sprigs

Bean sprouts

1 lime, quartered
Hoisin sauce
Sriracha sauce

Heat a cast-iron pan over a high flame until nearly smoking. Place the shallots and onions cut side down in the pan, and add the ginger around them. Cook, turning once, until all three are well blackened and blistered. Remove from the heat and set them aside.

Place the soup bones in a large Dutch oven or stockpot and add 16 cups of water. Bring to a vigorous boil, skimming off any foam. Add the carrots, peppercorns, blackened shallots, onions, star anise, cinnamon, cloves, coriander, and sugar, and bring to a boil. Reduce the heat to a simmer, and cook at a bare simmer for 2 hours.

Heat 1 tablespoon of oil or lard in a heavy skillet, and add the remaining shallot. Fry until the shallot is browned and sweet. Remove from the pan and set it aside.

Add the stew meat to the stockpot, and cook for another 1–2 hours, or until the meat is tender and the broth is richly fragrant and slightly reduced. Remove the stew meat and lightly shred it. Strain the broth into a clean pot. Discard all vegetables, aromatics, and soup bones (freeze those to keep urban dogs happy while you're at work). Stir the fish sauce into the broth, and bring it back to a brisk simmer.

While the stew meat is cooking, place the rice sticks in a bowl, and cover them with plenty of cold water. Allow to soak for at least 30 minutes. Bring a large pot of water to a rolling boil.

In each of 6 bowls, place a bit of shredded stew meat and a few slices of raw flank steak. Drain the rice sticks, and add them to the boiling water. Cook them for 30 seconds to 1 minute, then drain. Divide the noodles among the bowls, ladle simmering broth over the noodles, and garnish each bowl with fried shallot. Serve immediately. For flavorings and condiments, place the herb sprigs, bean sprouts, and lime wedges on a large platter in the middle of the table, and make the hoisin and Sriracha sauces available on the side.

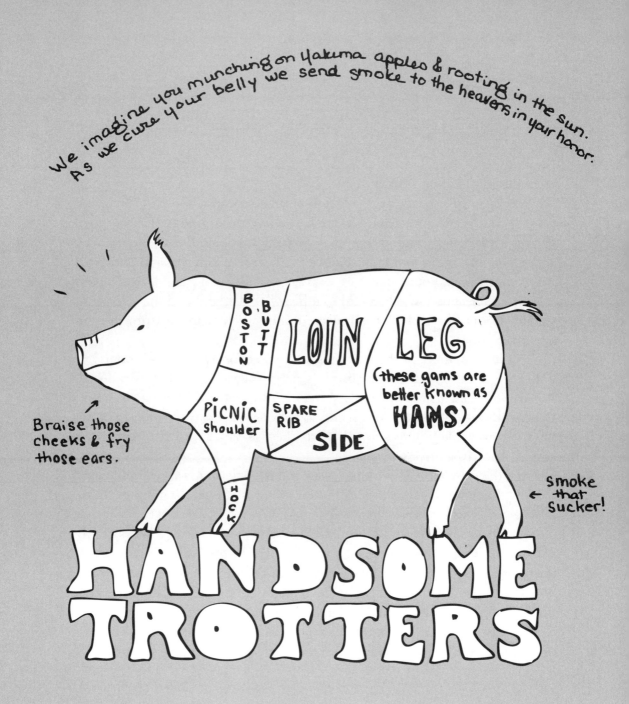

7

. . . .

PIG

If you have the space, money, and inclination, buying a whole or half-hog can be a wonderful experience for both novice and experienced cooks, as well as a splendid way to feed your family. Pork chops, roasts, bacon, and sausage are well incorporated into the American diet, translating into familiarity for the cook and a low "weird factor" for picky eaters and children.

The prodigious ground meat that a whole hog can deliver is quite easily made into sausage, offering more flavor variation and a way to work pork into breakfast as well as dinner. And as a wonderful bonus, pigs have versatile fat that can enhance your pies or preserve your meat. Advanced foodies can play at curing a whole fresh ham (the hind leg), pickling the trotters (feet), or making their own mortadella (an emulsion sausage laced with bits of fat). No matter what you do with your pork, its taste will be revelatory compared with that of the commercial product at your local supermarket.

To be clear, let's start by defining our terms. When I talk about buying a whole pig, I don't mean a suckling pig, the kind you bury in a pit in the ground and serve to a big group of people. That's what I call a single-use pig, because you really only want it for a day, though you might eat leftovers for a week.

No, I'm talking about a *hog*. One that weighs 250 pounds as it roots around and about 180 pounds when it's hanging on a hook. Like marriage, this is not a contract to enter into lightly—that's some serious pig, and we're not just talking bacon. In fact, we're talking about relatively little bacon overall. Just as a cow is not made up mostly of New York steaks, pigs are not made mostly of bacon. In fact, pigs contain no bacon at all. They do possess tasty "sides," which are divided into spareribs and a pork belly that can be cured or smoked to produce bacon. But a pork belly only weighs about 14–18 pounds, which means you have many, many more parts and a lot more pork to consider.

Lucky you! There is ground pork, seasoned ground pork (sausage), tenderloin, steaks, pork chops, cured ham, fresh ham, ribs, shoulder roasts...not to mention hocks, fat (lard), trotters, and I still haven't named all the goodies that will or could come from your hog. If what you are really after is just some home-cured bacon or a whole shoulder to roast for a Cuban-style pig party, there are easier and less expensive ways to go about it. Buy those cuts from your preferred producer or at a farmers market, or order from a reputable butcher. But if you are still undaunted and, like me, are giggling with glee at the idea of so much pork at your disposal, then the information below will help you decide between a half- or whole hog, and help you guide the processor into cutting the meat in a manner that works best for you and your family.

What Kind of Pig Can I Get?

While easier to pin down than beef, pork is still described in a number of different ways, from "natural" to "pastured" to "heritage" to "organic." One of the most confusing aspects of these varied descriptions is that at least one of these terms pertains to the *breed* of the pig, while others are related to the pig's feed or care. The definitions below will give you an idea of what to look for, but make sure to confirm these definitions with your grower.

Heritage pork. Pigs with pure heirloom bloodlines, or heirloom breed crosses, are sold as *heritage pork*. These breeds generally skew toward lard-type pigs and may take longer to reach market weight, causing the meat to be more expensive. Certain heritage pigs are specifically bred to forage and thrive in a particular environment or on certain types of feed, to produce more richly flavored and richer meat. I go into the breed question in more detail in the following section, "What Breed of Pig Should I Choose?"

Pastured pork. In comparison with pigs raised in hog-confinement plants, pastured pigs are allowed to freely forage, root, and roam in an open environment and do what pigs do. Their diet could include grass, brush, and trees in addition to nuts, legumes, dairy by-products, fruit, or grains. Because of its diet, pastured pork is reputed to enjoy a better nutritional profile, as with other pastured animals. Though pastured pork is often less lean than conventional pork, the increased fat is often *better* fat. Pastured pigs are also not given the low doses of antibiotics administered to conventionally raised pigs to combat the increased disease and stress caused by their factory environment.

Organic. In my experience it is relatively easy to find pastured pork and more difficult to find certified organic pork from smaller producers. When talking with growers, ask them which of the organic standards they follow and, if they're not certified, their reasons for not seeking certification. You may find, for example, that the animals are pastured in fields that are not sprayed with pesticides, and that might be good enough for you. Or the animals might not be dosed with antibiotics regularly but receive them when they are ill. This is a great opportunity to start a dialogue with your growers about their methods and communicate your consumer preferences.

Natural. According to the USDA, "natural" pork may not include artificial ingredients of any kind, including artificial color, and the product must be minimally processed. As with cows, I find growers use this word in a variety of ways in marketing their pork, in some cases as a synonym for "pastured," or to indicate pigs raised outside of a factory-farm setting. If your growers use this term, you'll need to ask precisely what they mean.

Should I Care What My Pig Eats?

When Cousin Kate and I went in on a pig that was raised by a farmer she knew, I asked what the pig was fed. She said he fed his pigs barley that he grew himself. I thought this sounded splendid. Tasting the meat, I was very happy. It was lovely pork, and I praised the barley that grew it fat and sound. A few months later, I went to a picnic at Kate's house. "You said you liked the pig we got?" she asked, without her usual edge of triumphant rural wisdom that colors many of our conversations. I did, I agreed, I thought it was wonderful.

"*Wellll,*" she began.

"What?" Now I was suspicious. Had I been poisoned? "Tell me!"

The "shocking" truth was just that the farmer had run out of barley while raising the pigs, and had finished them on waste corn from the frozen food plant nearby. I was expecting something far more horrible. It was good to know the truth, but if anything that just seemed resourceful.

If the slop you're feeding your pigs comes from the conventional vegetable factory down the road, it is different by degrees from the organic yogurt by-products another grower I met was feeding his. One is organic, for one thing, and the other isn't, and dairy is preferable to corn for feed in general, say some hog growers. In general, corn and soy are what factory-farmed pigs eat, and one argument against this feed is that it results in higher polyunsaturated rather that monounsaturated fat levels in the finished pork.

But the other differences in how factory-farmed pigs are raised contribute just as much to the equation. Depending on where they are raised, pigs eat a wide variety of foods, and pastured pigs graze on plants, trees, nuts, and fruit, all of which affects the taste of the finished product. I found the flavor of the next pig I bought, a pastured fruit-eating pig, distinct from that of that first barley–corn pig.

Even Cousin Kate agrees that a pig's diet makes a difference in terms of final flavor. Early in my process, she spoke of a local dairy farmer who keeps pigs to eat his waste milk, producing sweet, delicate meat. Learn what you like and ask your grower specifically what his or her pigs eat—and don't.

As with cows, reap the benefits of dealing with an individual grower by asking him or her directly what range of foods your pig will eat, how the feed was grown or produced, and why they chose that feed. Also note the following:

» Non-commercially raised pork is generally higher in fat, with better flavor.
» Pastured pork from healthy animals is lower in saturated fat and higher in omega-3s; the fat is also more stable, making curing more successful.
» A pig's breed can affect marbling, texture, and taste.
» Where your pig is processed can affect texture and quality as much as where your pig was raised.

» Try before you buy if you can. As with cows, many producers sell smaller packages or bundles of pork so you can sample the flavor and tenderness before committing to a whole animal.

What Breed of Pig Should I Choose?

Cousin Kate buys a half-hog each August from a young student who has raised the pig for 4-H. When I asked her what breed of pig she prefers to buy, she just shrugged her shoulders. "It's just a pig," she said, smiling. "Maybe a Duroc. They're expensive because of how they were raised, but they aren't 'heritage breeds.'" (Here she made scare quotes in the air.) The heritage breeds she was referring to are just that: pigs from pure bloodlines going back hundreds of years, with special characteristics particular to the breed that certain pig growers are working to conserve. In addition to their pedigrees, heritage pigs have glorious, evocative names like Large Black, Gloucestershire Old Spot, Tamworth, Red Wattle, and Poland China, that sound to me a bit like street handles for neighborhood toughs.

Kate's good-natured eye-rolling when it came to heritage hogs brings up a valid point for consideration when sourcing your pig. If you are the type that grows heirloom Scarlet Runner beans specifically to accompany your hazelnut-grazed Berkshire pork, you will do the research and pay the additional price per pound—which might be substantial—to buy into a heritage hog. For others, Cousin Kate included, the particular nuances of the meat and the curing quality of the pig's fat don't rank in the face of convenience and sourcing the animal locally. Most likely, the pig will be a breed cross (perhaps even of heritage breeds) developed by the grower for disposition, climate adaptability, size and growth characteristics, foraging abilities, hardiness, mothering abilities and milk production, and even color (dark skin protects pigs from the sun). Additionally, keep in mind that the larger (older) your pig, the higher the fat-to-meat ratio is likely to be.

Is Heritage Pork Worth the Price?

I have to admit, I was curious. You read all the descriptions of deep, nutty flavors, of lusciously marbled unctuousness, of meat so incredible it would be a crime to cook it rather than just rub it all over your naked body. Are the claims true?

I will tell you that every whole pig I bought from a local grower tasted better than any pork I bought in a store. I will also tell you that it tasted best right after I bought it and less so the longer it stayed in the freezer. But how, I wondered, would a pastured Berkshire pig taste in a blind taste test against a plain old pig—not a grocery store pig, but one that didn't have papers and had spent a good deal of time in a barn? (I know. If it were a scientifically valid comparison I would have compared a heritage-breed pig against a cross-breed raised in the exact same conditions. But I'm not America's Test Kitchen. You get what you get.) Would the difference be enough to justify the hundreds of dollars' difference in cost?

Armed with four pork chops and a dozen eggs, I headed over to Ethan and Angela's house. Just taking them out of the butcher paper you could clearly tell them apart. They were different hues, the non-Berkshire actually darker. But that wasn't the most apparent difference. Right now it was the processor, not the producer, that was informing how we experienced the pork. The Berkshire cut was a textbook example of a bone-in loin chop. The cuts were clean, the chop well formed and lovely. The other looked like it had been gone after with a dull hacksaw, and fascia and connective tissue crossed the thick chop at odd angles. The meat was pale. Not so pretty, and as it turned out, not so fun to chew.

We sautéed each in a hot film of olive oil with no seasoning save plenty of salt and pepper. We let them rest, and we tasted them before we broke the yolks on our eggs. The Berkshire tasted better, no doubt. Not only was the texture better because of the better cut, but the flavor was better, too. I would not describe it as remotely nutty; in fact, it was sweet, well-rounded. Whether this has more to do with the pig running around in a pasture, what the pig ate, or who the pig's ancestors were, I guess I can't say. What I learned is that the processor can make a big difference in the final quality of the meat you bring home and your experience eating it. While I don't know if I'd buy only Berkshires, I'd sure buy a pig from that producer again.

Lard or Bacon? (Do I Have to Choose?)

For most of us, how a pig was raised and what it was fed are ultimately more important information than whether our bacon-to-be was a handsome Chester White or a rare Guinea Hog. I assure you that if you buy into a local, pastured hog you will notice a difference in flavor, texture, and marbling, regardless of the breed. As you begin to seek out a pig of your own, you should be aware that while there are many breeds and crosses, they all began as one of two types: lard pigs or bacon pigs.

Lard-type hogs were once the pig of choice, a result of American growers breeding early European stock to develop a deep-bodied pig with an affinity for corn, and a natural talent for translating that corn into fat. The lard pigs enjoyed early popularity because of their high fat yield, especially during the World War II era, when lard was used in weapons production. Characterized by stout bodies and short legs, they develop fat before muscle and convert feed less efficiently in general. They also breed at a later age and produce fewer piglets than bacon-type pigs. Early on, nearly all American breeds fell under this category, with the Yorkshire and the long-bodied Tamworth worthy exceptions.

Mid-twentieth century, the lard-type pig began to fall out of favor. The pigs' decline came on the heels of chemical compounds replacing lard in lubricants, and vegetable oils taking the place of lard in American kitchens. Gradually, the country's breeds came to reflect a preference for more muscle, leaner meat, and less fat; lard nearly became a dirty word. Few pure lard-type breeds still exist, though some growers are touting heritage breeds like the woolly rock-star Mangalitsa, a Hungarian breed that looks like a cross between a sheep and a pig and whose fat is prized, especially for cured meat products. (Mangalitsas are rare— a chef's breed right now—with a price tag to match.)

Bacon-type pigs reflect the preference for meat over fat in their longer, leaner bodies (long bellies for bacon) and well-muscled legs (hams). In order to develop their characteristic muscle, these pigs are generally fed high-protein foods like nuts, legumes, or dairy by-products. This translates into a pig that grows more slowly as that muscle develops, with marbling in the meat the last bit of fat a pig packs on.

Here is where it gets a little confusing for the average consumer. Heritage breeds that may have begun as lard-type hogs—like the Chester White or the Duroc-Jersey—were later bred for bacon-type characteristics to reflect evolving market demand. The Berkshire, for example, is generally leaner than a century ago but has retained certain characteristic flavors and a darker color to the meat.

For most of us looking around at different growers, the best course of action is to simply ask what breed or breed-cross they raise and what characteristics those pigs exhibit. Many small-scale pig growers feel passionately about their pigs, and if asked will wax poetic in much the same way as a proud Weimaraner owner will detail the breed's noble game-hunting history and the reason for its characteristic coat. If you are working through a processor who sources animals from local farms, you may also be able to buy a small amount of the pork, do a taste test, and make sure it's the pig of your dreams.

What Cuts of Pig Can I Order?

If you've just gotten through reading the Cow section, you'll be glad to know that breaking down a pig carcass into recognizable retail cuts is a much simpler process. Just as with a beef carcass, you begin by dividing the body into primals. On a hog, you only have half the amount of primals to remember as on a steer.

The four primals are the Shoulder, the Loin, the Side, and the Leg. Before we go into those, though, here's a little information about the trim from a hog carcass, which is a source of ground pork.

Unlike with beef, most processors will make fresh pork sausage for you. Some offer just a couple of options, such as sweet or hot Italian-style sausage, or Italian and breakfast, for example. Others go crazy on the sausage front and offer loads of options. Before you order sausage from your producer, buy some to try first if you can. Especially with processors that offer sixteen flavors, you can expect that they have sixteen different off-the-shelf "flavor powders" ready to mix with your pork, in a process more akin to making

Sno-cones than sausage. Some processors take more care with their sausages and smoke and cure their own products as well. Usually you'll receive better products from these folks. Asking for plain ground pork and fat gives you lots of options for making your own sausage, including sausage stuffed into casings.

Primal Needs

Let's begin at the front of the animal with the **Shoulder**. This primal makes up about one-quarter of the hanging carcass weight and delivers flavorful meat with fairly significant marbling and connective tissue, due to the pig's extensive use of its shoulders and front legs. This primal is further divided into the upper portion of the shoulder, counterintuitively known as the Boston butt, and the lower portion, festively called the picnic.

Shoulder roasts from top or bottom are ideal for slow braising, roasting, and smoking. This is the cut for Southern-style slow-cooked barbecue, for braising in milk, or for roasting with herbs. A cut rarely seen in grocery stores but both versatile and delicious, pork blade steaks are often cut from the shoulder and are great on the grill. And the shoulder primal isn't just about the shoulder. Don't forget about that little leg, or hock, down there. Processors will usually smoke the hocks for you, and they make an amazing addition to beans and greens. If you want the trotters (the hooves) as well, make sure you order them ahead of time.

Pork Shoulder Cuts:

Boston butt roast   Shoulder or blade steak
Arm picnic roast   Hock
Blade roast

Continuing back along the animal, we enter into pork-chop country when we arrive at the **Loin**. As with the cow, the loin consists of tender, fairly lean meat that must be cooked with care, even if it's from pigs with higher fat content. This primal makes up slightly less than a quarter of the carcass and contains the loin, the sirloin, and certain rib cuts. You can have the entire loin removed and cut into boneless chops, or have bone-in chops cut from the loin as well. The tenderloin is a smaller piece best left whole. If you order boneless chops you'll have the option of receiving baby back ribs. Country-style ribs hail from the loin as well, but need slow cooking like cuts from the shoulder.

Pork Loin Cuts:

Bone-in or boneless loin roast	Baby back ribs
Bone-in or boneless pork chop	Country-style spare ribs
Tenderloin	

Let's dip on down to the **Side**, aka the luscious pork belly, home of bacon, of pancetta, and of spare ribs, of course. It's just under a quarter of that carcass and it's mostly fat. This is the primal renowned for meat so well streaked with fat it will induce British accents as you say "Please bring me a rasher of streaky bacon!"

My bet is that your processor will be all too happy to cure and smoke your bacon for you, and slice it up. That's a fine idea. But you can also cure and smoke your own, or just rub it with herbs and pepper and cure it and roll it and trot out your own pancetta. You can also braise and roast your own pork belly and forget the bacon thing altogether. Truth be told, whatever you do with it, I don't think you can go wrong. You can barbecue the spare ribs, but you can also steam them in rice wine, soy sauce, and star anise, then broil them with a gingery glaze.

Pork Side Cuts:
Belly
Bacon
Spare ribs

Though I'd love to linger on the Side, let's finish up with the pig's prodigious thighs, which make up a full 30 percent of the carcass weight. The **Leg** is also known as the **Ham**, and if you took your standard cured ham, turned it upside down, and tilted it a little, you could see where it originally fit onto the animal. Just as with bacon, the processor will surely cure and smoke a ham for you. But if you've never had one before, I urge you to have the processor cut that leg in half so you receive two fresh ham roasts instead. They're delicious. If you've managed to get a pig with the skin on, so much the better, as the skin helps keep the hams moist.

If you do receive a skin-on pig, you can also endeavor to salt and cure the whole darn thing and turn it into a country ham or your own prosciutto. When I took a local butchering class, another attendee told us all that he was curing a prosciutto in his guest room. He thought it was going pretty well, though he said he has to move it to the garage when his Jewish in-laws come to stay. And for me, the chance of screwing up an entire leg of a beautiful pig is just too daunting... but once Husband Hendel builds our curing chamber, *Look out, world!*

Pork Leg (or Ham) Cuts:

Bone-in fresh ham	Cured ham
Bone-in fresh ham roast	Hock
Smoked ham	Trotter

Though it is not technically a primal, and even if you're not a lover of pig knuckles and would never think to ask for the cheeks, consider asking for the trimmed fat to make your own **Lard**. As we learned earlier in this section, there is evidence that pastured pig fat is not so bad for you. Though I wouldn't say its health effects are tantamount to eating nothing but flax seed and spinach, it might be good for you in the same way that chocolate, red wine, and coffee are. (Maybe.) And it's delicious. I would like to try curing my pig's lard into a product the Italians call *lardo*, but I haven't done it yet.

Leaf lard is the "it" fat these days, and people will sing its praises to the moon, saying it's best for pastries. This is the fat that surrounds the pig's kidneys. It is reputed to be of the highest quality, imparting the least amount of porky taste to foods baked with it, as well as a crystalline structure that creates shatteringly crisp pie crusts and cookies. I will say that back fat—the thick layer of fat that runs along the hog's back—is easy to trim and clean.

I have found that either one works well for baking and frying, if you take care not to let any particles brown in the lard as it renders. Also, pork fat is a common ingredient in sausages made of leaner meats, like elk, deer, even lamb. The sky's the limit!

Tips for Success

If you are ordering a half-hog rather than a whole one, know that your choices will necessarily be more limited. One cheek or ear only goes so far. You cannot ask for the whole belly. It is in this instance that you will reap the most benefit from going in with family or friends and ordering larger pieces from the producer. You will have more flexibility to get the pork cuts you want if you take the whole animal. Also note:

» Give the producer and processor plenty of lead time if you have special requests, like wanting the head, cheeks, trotters, or innards. You may not be able to get the head, tripe, tongue, ears, or other cuts even if you ask. Be prepared.
» If giving cutting instructions still scares you, ask the producer to guide you in making requests that reflect the way you cook. If you simply adore sausage, for example, then ask for more trim. If you want the easiest route, have the processor smoke and cure everything they can—from the hocks to the hams—*and* make sausage for you.

How Much Will My Pig Cost and How Much Will I Get?

First, you need to decide whether you want the whole animal or a half, which is an option for most producers. Because of the size of the animal, you cannot buy less than half a hog directly from a grower (though there is nothing to stop you from dividing it up once the pork is in your possession). As with most whole meat, pork pricing is usually based on hanging weight. And as with beef, you will usually be asked to pay a deposit to reserve your animal.

In general, expect to pay around $3.50–$4 per pound including processing for a pig when ordering the whole animal.

As we did with the cow, let's run through two examples, one based on an inclusive price paid directly to the producer and one in which the producer and processor are paid separately for their services. Each scenario is based on actual fees charged by small-scale producers, one here on the West Coast and one in the Southeast:

Spotted Pig Farms charges $3.99 per pound for whole pigs, including slaughter, processing, and smokehouse fees. They give an average hanging weight of 180 pounds.

(average hanging weight) 180 x
(inclusive price per pound) $3.99 = $718.20

Producer Happy Pig charges $2 per pound for a whole animal, but let's say you can't find anyone to go in on the pork and your freezer is just too small. We see that they charge $2.50 per pound for a half-pig, excluding the kill fee and cut-and-wrap. On their website, the average hanging weight of a half-hog is listed as 90 pounds. Unless you specify a different processor within a 50-mile radius of their farm, they work with their local processor, Acornfed Meats. Acornfed Meats charges $.45 per pound for processing, with an additional $55 disposal and kill charge per order. They will smoke products for an additional charge of $.75 per pound.

You have to calculate two costs here and total them together. The first is from the grower for the meat.

(average hanging weight) 90 x (price per pound) $2.50 = $225

You also need to pay Acornfed Meats to slaughter the animal, cut and wrap it, and smoke your bacon and ham hocks. Let's estimate the weight of those at about 15 pounds total:

(90 x $0.45) + (15 x $0.75) + $55 = $106.75
$225 + $106.75 = $331.75 total cost for a half-hog

If you divide the first cost in half, you will see that the two are nearly identical when all is said and done. Keep in mind that these costs are for average pastured pigs. Expect to pay substantially more for heritage breeds and for animals that are certified humane and organic.

Questions to Ask Your Grower:

» Is the cost calculated by live weight or hanging weight?

» How much deposit is required?

» What breed are your pigs? Why do you raise that breed or cross?

» What is the average hanging weight of your animals?

» Are slaughter and processing included in the fee, or will I pay the processor separately?

» Will the pig be scalded and scraped, or skinned? (See more about this in "Who Will Kill My Pig and How?")

How Much Meat Will I Get?

Some growers charge more per pound if you order only a half-hog, as they then must match you up with another customer who wants the other half of a similarly sized animal. This might also change the timing of when you receive your meat. Small-scale growers, especially, cannot simply produce a pig for you overnight. Finding your own pig partner and divvying up the meat between you can be a fun and money-saving option.

Sharing a hog also allows you to do a little horse-trading, so to speak, so that you end up with the cuts that work best for you and your family. For example, an average whole hog contains two fresh hams (back legs), each weighing approximately 15–18 pounds. Dividing each fresh ham in half into the butt and shank reduces the size of your roasts, but doesn't change the amount of meat you must fit into your freezer and either cook for a crowd (*see* Coke Ham in Recipes) or eat for days.

If you are less excited about ham but would happily eat sausage morning, noon, and night, imagine the love connection you could forge with that crazy friend who craves two fresh ham roasts plus a whole leg to cure and smoke. In exchange, he slides you over his share of ground pork, and everyone is happy. Buying into a whole hog with someone you know also means someone can receive the entire head, or two ears or cheeks, an amount of meat you can actually work with.

As for how much meat you will take home, I have found less variability in the slaughter weight of pigs than in cows or lambs, with most hovering around 250 pounds on the hoof and some going down to 200. In general, pigs lose less weight in the translation from animal to packaged meat, with a dressing percentage of about 72 percent. This means that the hanging weight of the carcass will be about 180 pounds on average. Of that carcass, from 65 to 74 percent will be recouped in the amount of take-home pork, depending on your cutting instructions and on your individual hog. Obviously, the more offal, fat, and bone you take, the greater that percentage will be. Let's look at the range in pounds of take-home meat that an average hog will deliver:

180 pounds hanging weight x .65 = 117 pounds of take-home pork
180 pounds hanging weight x .74 = 133 pounds of take-home pork

In order to guesstimate your take-home amount of meat, ask your producer for the average hanging weight of their pigs and then apply the percentage appropriate to the cuts you are likely to order—bone-in or boneless, closely trimmed or fat on, organ meats and trotters included or not, and so on.

If you went with "off the shelf" cuts and didn't ask for any extras like fat or offal, here is what a whole pork would look like as recommended by a grower I've bought from:

48 pork chops	4 smoked hams
16 pork steaks	14–16 pounds of bacon
4 shoulder roasts	8 hocks (they are usually cut in half)
2 loin roasts	6–8 pounds of sausage or ground pork

How Much Freezer Space Do I Need?

Half a hog will completely fill the freezer compartment of an average refrigerator/freezer unit, while a whole hog will occupy half a standing freezer unit or one-third of a larger chest freezer.

Who Will Kill My Pig and How?

You should care how your pig is killed, and not just because you want to stay in the black in terms of karmic debt. Especially with pork (though this is true of other meats as well), undue stress at the time of slaughter can result in pale, soft, exudative—called PSE—pork. As the name so aptly describes, the flesh is very light-colored, has a mushy texture, and gives off liquid as it sits in its packaging, resulting in drier cooked pork. Certain pigs have a genetic predisposition to developing porcine stress syndrome, or PSS, which can cause PSE, especially in pigs bred to put on muscle as quickly as possible. Temperature is also a factor, as PSE is more of a risk due to improper chilling of the carcass and during warm weather.

When pigs are stressed by being denied access to food and water, for example, or traveling immediately prior to slaughter, or being prodded or poked, or seeing other animals being slaughtered, they get excited, in the wrong way. Even flashing objects and shadows can stress them. Stressed pigs have higher lactate levels in their muscles, which contributes to an abrupt drop in pH that causes PSE. Calm environments and on-farm slaughter done by professional butchers can be more humane and results in higher-quality meat overall. While pigs raised in confinement operations are rendered "insensible" either by being gassed with CO_2 or through electric stunning, your small-producer pig is more likely to be shot in the head, or stunned with a captive-bolt stunner, to render the animal insensible before its throat is cut to bleed out. Or the processor may just stick the pig without stunning.

One important influence on the pork you receive comes into play after the animal is killed. Following slaughter, pigs are either skinned, like beef or lamb, or scalded by dipping them in boiling water, followed by the hair being scraped off the hide. Skin-on pork is preferred by many people and is a requirement for pork that will be cured in certain ways. Ask how your pig will be killed and processed before you buy, as some growers only work with processors that skin hogs, making skin-on pork unavailable. Other growers will only provide you with skin-on pork that is fresh; having it processed and frozen will cost extra.

When Will It Be Ready?
Unlike beef, processors don't age pork. If you order smoked or processed products—including hams, hocks, bacon, or sausage—you may need to allot a bit of extra time for the smoking or curing. From slaughter to pickup averages about a week, I've found, more if you want the hog to come to you rather than the other way around.

RECIPES FOR PIG

Workday Coffee-Aleppo-Rubbed Pork Roast

It could be a Seattle thing, but I take coffee any way I can get it. Here it adds an earthiness, along with the ancho, to the rub, taking the pork uptown for a little ride. I made this when I had some Aleppo pepper on hand from a visit to Penzeys Spices. This Syrian pepper has a mild shot of heat and a rich fruity flavor that's much more complex than that of chile flakes. And it's mild enough that my kids will eat it. If you don't have Aleppo pepper, substitute whatever mixture of ground chiles makes you happy. This recipe makes about twice as much rub as you'll need. Use half for rubbing the roast, and place the rest in a tightly sealed container. It will keep for up to three months and is delicious on ribs, too. Also, for you slow-cooker fanatics, this is an excellent candidate.

SERVES 4–6

2 tablespoons Aleppo pepper
1 tablespoon ancho chile powder
4 teaspoons salt
4 teaspoons finely ground coffee
2 tablespoons brown sugar
½ teaspoon onion powder
½ teaspoon garlic powder
Small pinch ground cloves
4 pounds pork shoulder blade roast

Combine the Aleppo pepper, ancho chile powder, salt, coffee, brown sugar, onion and garlic powders, and clove in a small bowl. Set aside half the mixture for another use. Heavily coat the roast with the remaining mixture, rubbing it into the meat and covering all surfaces. This can be done the night before (refrigerate it overnight) or the morning of the day you want to serve the roast.

Preheat the oven to 200°F.

When the oven is heated, place the roast in a Dutch oven, cover with a lid, and place in the oven. Cook for 6–8 hours, or until the meat is falling off the bone. The low temperature gives you a lot of leeway in terms of time. If you'd like to develop a little crust, raise the heat to 300°F, and remove the lid for the last half-hour.

Slice the meat thickly and serve, spooning accumulated juices over it. A little pillow of grits under there would be nice. Steam any leftover pork the next day with a few tablespoons of apple cider vinegar, then shred and stuff it into rolls for pulled pork sandwiches. Pass the coleslaw.

On Bacon Fat and Ends

○ ○ ○ ○ ○ ○ ○ ○ ○ ○ ○ ○ ○ ○ ○ ○ ○ ○ ○ ○ ○ ○ ○ ○ ○

I trust, since you are reading this book, that you are not the type of person who considers turkey bacon real food. The whole point of bacon is the fat; if you can't get over that, I can't help you. Save that fat! Fry eggs in it, deglaze with vinegar and whisk in some mustard and pour it over sturdy bitter greens, use it for sautéing fresh corn kernels…you don't need me for this. Keep a Mason jar in your fridge and let me know what new uses you find. And ignore your husband when he mocks your "farm wife ways"—he should be so lucky.

Bacon ends are now sold, packaged, in some grocery stores, but they are inevitable when you order half- or whole hogs and have the bacon smoked for you. As it turns out, hogs do not grow perfectly rectangular bellies that slice ever so neatly. The knobby bits cut off when squaring the belly to deliver you that picture-perfect slice—those are bacon ends. The nice part about ends is that they are much simpler to cut into any shape you desire, from a true dice to *lardons*.

Bacon, Farro, Potato, and Kale Soup

Making dinner on weekdays in general elicits in me an emotion far from joy. There is the rushing home from work, the picking up of kids from aftercare or kung fu, the neighborhood or school meeting right after, or during, dinnertime. So on the weekends, when days seem to stretch luxuriously before me and keeping pots on the stove for eight hours seems less a fire hazard than an homage to simpler times, I like to get my cooking on. Best are the meals that braise or stew or simmer untended and provide dinner two nights in a row. This soup's broth began as a way of quick-thawing a pound of bacon to fry for Sunday breakfast. I thought if I steamed it just a little I could pry the slices apart more easily and lose some of the fat. Well, it worked like a charm, but then I repurposed all of that fat and flavor in a soup. Wait, did I do something bad?

SERVES 4

1 pound frozen thick-cut smoked bacon, not too lean

Olive oil for frying and garnish

1 yellow onion, diced

3 cloves garlic, thinly sliced

1 cup farro

1 pound new potatoes, peeled or scrubbed and cut in large dice

½ head Tuscan or other kale, deribbed and cut into ribbons

Place the bacon and 2 cups of water in a pot with a tight-fitting lid. Bring to a gentle simmer and cook for ten minutes, or until the slices fall into the water. Reserve 4 slices and the bacon water, separately. Remove the rest of the bacon and fry it for breakfast, so the children will let you continue to make soup.

Dice the 4 parboiled slices, or cut into thick ribbons. Fry them in a little olive oil in a Dutch oven or heavy pot until lightly golden. Add the onion and garlic and cook, stirring frequently, until the onion is soft. Add the farro and cook, stirring, just to coat the grains with fat.

Add the potatoes, bacon water, and enough additional water to total 1½ quarts, along with a few pinches of salt. Simmer for 20–30 minutes, or until both the farro and the potatoes are tender. Taste for salt. Add the kale ribbons, and cook for an additional 5 minutes, or just until they soften.

Serve the soup with a dash of olive oil and plenty of freshly cracked black pepper. I don't think a thick shaving of Parmesan floating on top would be objectionable, either.

Caribbean Pork Sandwiches

Husband Hendel believes that the pork sandwiches from Paseo, a North Seattle Caribbean restaurant, are hands down the best sandwiches on Earth. Though I go back and forth between the pork and the shrimp, there is no denying that dripping olive oil, luscious mayonnaise, deeply caramelized onion, and enough chopped garlic to make you faint makes for a very, very fine sandwich, and one you can make at home. These are in no small part about the mayonnaise, so make your own. And use cilantro like lettuce. And caramelize the onions, really caramelize them.

Note: The easiest way to pound the meat is to slit open a resealable bag and sprinkle the inside with water. Center the meat on the bottom flap, cover with the top flap, and then pound with steady beats beginning in the center and moving outward. I'm sure it won't surprise you to learn that children love to pound meat. If you let them, get over wanting it to be even—too even, anyway. Fun is always more important than appearances.

SERVES 4

⅓ cup olive oil, plus more for frying
5 juicy cloves garlic, minced
¼ cup fresh orange or tangerine juice
¼ cup freshly squeezed lime juice
Pinch ground cumin
1 whole pork tenderloin, cut into quarters and then pounded thin, or 4 boneless loin chops, pounded thin
3 large yellow onions
1 large egg yolk
1½ teaspoons Dijon mustard
3 tablespoons lemon juice
Pinch of salt
½ cup liquid bacon fat

1 cup canola oil

Small bunch cilantro with stems, washed and well dried

4 long, crusty rolls or one baguette, ends trimmed and cut into quarters

At least 1, and up to 4, hours before you want to eat, prepare the marinade. Heat the ⅓ cup olive oil over medium heat, then whisk in the garlic. Immediately pour the mixture into a small dish, then add the orange (or tangerine) juice, lime juice, cumin, and salt to taste. Add the pork and toss to coat, then place the meat and marinade in a resealable bag. If you're marinating for only an hour, leave the pork out on the counter. Otherwise, refrigerate it. Turn the bag every now and again.

Slice the onions thickly in rings, at least ½ inch thick. Heat a large sauté pan or Dutch oven over medium heat, and add a glug of olive oil. Add the onions, sprinking liberally with salt and just a pinch of sugar if you like. Toss the onions to coat. Cover the pan and cook over medium heat for 10 minutes just to get the onions going, then remove the lid and cook over low heat until the onions are deeply browned and very soft, about 45 minutes. Set aside.

Make the mayonnaise by placing the egg yolk, mustard, lemon juice, and salt in a blender or food processor. Pulse or blend to combine. With the motor running, add the bacon fat and canola oil in a thin, steady stream. Taste for seasoning.

Heat a large sauté pan over high heat. Film the bottom with olive oil. Remove the pork from the marinade and pat it dry. (It's not necessary or even desirable to remove all the garlic bits.) Place the pork in the hot pan, and and sauté just until brown on each side, 1 to 2 minutes per side. Slice the bread lengthwise, and slather each side generously with mayo. Add a bed of cilantro sprigs, and place the meat on top. Cover the meat with caramelized onions and serve.

Grilled Ham Steak
with Balsamic Peaches

The day Ingrid and I wrestled that first pig home turned out downright hot. The joys of refrigeration mean we don't have to wait for fall anymore to buy our pigs. I was so excited to taste my purchase that I defrosted a package of pork steaks immediately and threw them on the grill. The steak is a cross-section cut, in this case from the shoulder, and it's made for the BBQ. Nestled alongside grilled Yakima peaches with a dash of balsamic vinegar, it was a meal, and as we sat on the patio we got to enjoy the last rays of the setting sun.

SERVES 4

2 large, ripe freestone peaches, halved and pitted
Olive oil
Kosher salt and black pepper
4 whole sage leaves, washed and dried
½ cup balsamic vinegar
4 pork shoulder steaks, cut 1 inch thick

Toss the peach halves in olive oil to coat, and sprinkle with kosher salt and cracked black pepper. Press one sage leaf against the cut side of each half. Set the peaches aside.

Heat the balsamic vinegar in a small saucepan over high heat. Reduce it by half, about 5–8 minutes.

Heat a gas grill to high, or light a charcoal grill and wait for the coals to ash over.

Season the pork steaks on both sides with salt and freshly ground pepper. Grill them for for about 6 minutes on each side, turning them only once. (Reduce the grilling time if your steaks are thinner.) Loosely cover the steaks, and set aside to rest.

Over medium heat or at the edge of the coals, grill the peaches cut-side-up until marks form. Carefully turn and grill the other sides just until the fruit is heated through. You should be able to smell the sage, and the fruit should be hot but not mushy.

Nestle one peach half, sage side up, next to each steak on the plate. Lash each peach with the reduced vinegar and serve.

Very Good (Cuban) Black Beans

Seattle doesn't have a big Cuban community, which is sad, because Cuban black beans are as close to inky silk as you can get, with a perfect finishing touch of sugar and vinegar. When I was a vegetarian I was as thrilled as could be to eat black beans and rice for dinner, hoping no pork was used in their making but adopting a realistic don't-ask-don't-tell policy most of the time. Now, I welcome the pig fat and the smoked ham hock.

You can use either bacon drippings or lard to sauté the sofrito. I use lard, since I find the hocks impart enough flavor on their own; if you want to smoke it up, go the drippings route. And if you happen to have just rendered your lard and have some chicharrones (fried pork rinds) on hand, crumble them on top. These beans freeze beautifully, or leftovers make a lovely lunch the next day.

SERVES 6

1 pound black beans, picked through and soaked overnight

2 bay leaves

2 yellow onions

1 large ham hock, or two smaller chunks, depending on the cut

1 or 2 habañero chiles, pierced with a fork (optional)

3 cloves garlic

Sprig of oregano

2 tablespoons lard or bacon drippings

2 green bell peppers, diced

Kosher salt

Brown sugar

Red wine vinegar

Diced firm banana, for garnish

Drain the beans and place them in a large Dutch oven. Cover them with 1½ inches of water. Add the bay leaves and one of the onions, halved. Nestle the ham hock in the beans. If you want it hot, add the habañero or two. Bring to a boil, then reduce the heat to a simmer and cook until the beans are just tender. This will probably take an hour, but keep checking. Once the beans are soft, add salt if the hock hasn't added enough. Taste to be sure. The liquid should be reduced and satiny.

While the beans are cooking, make the sofrito. Dice the remaining onion and chop the garlic. Remove the oregano leaves from the stem and roughly chop them. In a sauté pan, heat the lard or drippings over medium-high heat. Add the onion, garlic, and peppers and a good sprinkling of kosher salt. Throw the oregano on top. Sauté until the vegetables are soft but not brown.

Remove the onion and bay leaves from the beans (and compost them). Take out the hock, and use forks to shred the meat—save the bones for the hounds, if you like. Reserve the meat.

Add the vegetables to the beans and stir. Using an immersion blender, purée the mixture until it is silky but there are still whole beans and vegetables. Add the shredded meat back to the pot. Cover, and cook on low for another 15 minutes to let everything meld.

Just before serving, add the brown sugar and vinegar. This is a personal thing, but start small—a couple of teaspoons of sugar to begin, a little slosh of vinegar. Stir it and taste it, and add more if you like. Remove the habañeros so no one dies. Sprinkle the top with a little banana, fried or not, and serve.

On Smokers

○ ○ ○ ○ ○ ○ ○ ○ ○ ○ ○ ○ ○ ○ ○ ○ ○ ○ ○ ○ ○ ○ ○ ○

I have to tell you, it was not just the promise of a tax write-off that prompted me to buy the smoker. For one, I like smoked foods, and I was sorely disappointed when our Little Chief disintegrated outside under the pine tree, even though I found its smoke level widely variable and hard to control.

Second, Husband Hendel did not argue with the purchase of a smoker, because he too likes smoked foods. Unlike an auxiliary KitchenAid, for example, or ten sizes of cake pans, or an immersion blender, he could see the direct benefit our having a good smoker might have on his life—smoked fish, smoked cheeses, smoked meats—it's all good. He even *used* the smoker (see the "Smoked Pork Chops with Fried Green Tomatoes" recipe).

Thirdly, though not lastly, ALL of my brothers have smokers. They have expensive electric ones that use little smoked wood pucks with old-fashioned ones on the side. Dee has an electric one AND his handmade refrigerator smoker that could handle a whole moose at a time. I'm not saying it's a contest, but I have a reputation to protect. It is for all these reasons, plus the whole animals living with me, that the smoker just made sense.

As Husband Hendel can testify, buying whole animals makes your inner cook go a little frontier. Suddenly you need dehydrators, smokers, canning pots, and freezers. Jerky slicers, meat slicers. . . . I even looked at a field dressing table. Who was I becoming?

Truth be told, the smoker was one of our finest investments, and here's why:

» **Smokers have come a long way.** My smoker is electric, with programmable temperature controls, and uses little compressed wood pucks in the place of wood chips. No soaking! I can cold-smoke foods, keeping the temperature under 90°, just to flavor them if I go on to cook them later like pastrami. I can also hot-smoke them, usually around 225°, to both flavor and cook the meat. Fancier ones than mine let you monitor the smoke level and temperature of the food from inside the house.

» **It is an appliance of urban size.** It sits, under a pretty little cover, on my front porch right by the porch swing. Now, don't go taking it if you know where I live—if you just ask nicely, I will smoke something for you.

» **It will help you preserve your meat.** Not only does smoking meats into sausages, jerky, pastrami, and Southern-style BBQ offer variety and deliciousness. In some cases it allows you to preserve your meat and extend its shelf life. Like pemmican, but better.

» **It's not just good for meat.** I do love smoked sausages and pork chops and pastrami and . . . you get the idea. But do you know what I love more? Smoked mackerel. Holy moose meat, that's good! And incredibly easy. Smoke cheese and vegetables, make your own chipotles. Trust me, you'll be happy.

Bacon Jam

My friend Valenti loves bacon more than anyone I know. She eats chocolate-covered bacon candy, has bacon cookbooks, and can smell it cooking from three miles away, upwind. She's so obsessed that for her birthday once I endeavored to create a bacon-themed dinner, from candied bacon–adorned Manhattans to red-cooked pork belly to honey-drizzled doughnuts fried in lard. So when she said she had mastered her own version of bacon jam, I knew I had to have that recipe. I tasted it, and loved it so much I began to understand Valenti's unwavering devotion. It is unctuous without being overly rich, and just sweet enough, with deep undertones of coffee and smoked paprika. Eat it with tart apples and a baguette, make it part of a cheese tray, slather it on a grass-fed burger, or just dip into it with your finger. Making it with your own bacon puts it completely over the top.

MAKES 1 PINT

1 pound thick-cut good-quality bacon, cut into 2-inch pieces

1 medium yellow onion, chopped

2 shallots, minced

4 cloves garlic, minced

¼ teaspoon chipotle chile powder (adjust according to personal taste)

¼ teaspoon smoked paprika

¼ teaspoon ground allspice

¼ teaspoon ground cinnamon

1½ cups strong brewed coffee

⅔ cup apple cider vinegar

⅔ cup (packed) brown sugar

½ cup pure maple syrup

In a large, heavy pot, cook the bacon over medium-high heat until browned and crisp, stirring occasionally. Drain the bacon on paper towels, and remove all but about 2 tablespoons of rendered fat from the pot. (I *know* I don't have to tell you to save that fat, do I?)

Add the onion and shallots to the pot with the bacon fat, and cook over medium heat until soft and translucent, about 10–12 minutes. Add the garlic and spices, and cook until fragrant, about 1 minute.

Deglaze the pan with the coffee and apple cider vinegar, stirring to scrape all the bits from the bottom. Return the bacon to the pan, and add the brown sugar and maple syrup. Bring to a boil, then reduce the heat to a simmer.

Simmer over low heat, stirring every now and then to prevent sticking, for 3–4 hours, or until it smells so delicious that you can't wait any longer. The jam will be a deep, rich brown color and have a jammy consistency, more gooey than liquid-y.

Let the jam cool, and transfer it to a food processor. Pulse a few times to finely chop and blend the jam.

Valenti would eat it right from the bowl of the food processor with her hands. But if you have a shred of self-control you should place it in a glass jar and refrigerate to allow the flavors to meld. The jam improves with time and will last a couple of weeks in the fridge.

Pressure Cooker Carnitas

During the ill-timed kitchen remodel, I took the kids and escaped Seattle to the company of my brothers and their intact kitchens—luxurious always but heaven for someone cooking on a hot plate. I found Erik measuring and prepping for this intriguing dish, and I investigated the pressure cooker on his counter with some interest. I have to admit, I'm scared of them. I still remember all those stories from my vegetarian days of beans exploding in people's faces. But I hear that things have changed, and that I am a sissy for worrying about them anymore. Whatever my prejudices, there is no denying that Erik started dinner at four, and by six, when the rest of the brothers three arrived with toddlers and fresh tortillas in tow, it was ready. I mean, we were sitting around drinking margaritas already. And the meat was divine. Fast and good? I'm in.

SERVES 6–8

    Olive oil for frying
    4 pounds pork shoulder or country-style ribs
    Juice and zest of 1 orange
    Juice and zest of one lime, plus juice of a second lime
    2 cups chicken or pork stock

2 tablespoons chili powder
1½ teaspoons dried oregano
1 teaspoon cumin
2 jalapeños, sliced
5 garlic cloves, sliced
Chopped onions
Cilantro

Film a saucepan with olive oil, and heat on high. Brown pork well on all sides.

Whisk together the juices, zest, stock, chili powder, oregano, cumin, jalapeños, and garlic. Place the pork in a pressure cooker, and pour the mixture over it. Cook on high pressure for 75 minutes.

Remove the meat to a plate, and shred it with two forks or a sharp knife. Strain the liquid through a sieve into a saucepan, to catch the garlic and jalapeños, and reduce by half over high heat.

Cover the bottom of a heavy nonstick skillet with oil or lard, and heat over medium-high. Fry the shredded meat in batches until lightly crisped. Add a couple of spoonfuls of the liquid, and cook until the meat is glazed and all the liquid has evaporated. Serve with chopped onions and cilantro with fresh corn tortillas.

Coke Ham

With apologies to my small-town Lutheran roots, I have to confess that I don't really like ham. I'm not talking about Smithfield ham, or prosciutto, but the kind that comes adorned with cloves and canned pineapple rings and is a must at Easter brunches and funerals. It's just not my thing. But unless you shop at an ethnic market, that is probably the only treatment of the pig's hind leg you will find. And that is sad, because fresh ham roasts put the more common loin roasts to shame—they're juicier, more flavorful, and they're just plain delicious. Way before I knew what a fresh ham was, I read a recipe in Cook's Illustrated *that was based on an old Southern tradition of marinating pork in Coca-Cola. Intrigued, I ordered a fresh ham from the butcher and, following the recipe to the letter, was more than pleased to produce an enormous hunk of juicy meat that was sweet and salty and herbaceous, and just as amazing cold* the next day. I've made Coke ham many, many times since then, experimenting with different rubs and glazes, trying to perfect my cracklings. When I ordered my first half-pork, I of course asked for a fresh leg, cut in half. (Pigs have really big thighs, and you don't want to have to go at that thing with a hacksaw. The voice of experience here.) Because most whole or half-pigs come skinless, I've adapted the recipe to suit. It's still delicious without cracklings, though perhaps less Southern. Don't forget the biscuits. Dice and fry leftovers and tuck meat into tortillas.

SERVES 8–10

3 cups plus 1 tablespoon kosher salt

½ cup plus 1 tablespoon whole black peppercorns

2½ juicy heads garlic, peeled

1 fresh ham roast, sirloin or shank (half of a whole
    fresh ham or pig leg)

3 2-liter bottles Coca-Cola Classic

½ cup fresh sage leaves

½ bunch fresh parsley

2 teaspoons coriander seeds

¼ cup grapeseed oil

1½ cups light brown sugar

1 cup pineapple juice

¼ cup apple cider vinegar

2 bird's-eye chiles, minced, or pinch red pepper flakes

Place 1 cup of the salt and ½ cup of the peppercorns in a food processor. Add the cloves from two heads garlic. Pulse 3 to 4 times to coarsely chop the garlic and crack the peppercorns. Rub the pork all over with salt–garlic mixture, and allow it to sit while you prepare the brine.

Dissolve the remaining two cups salt in the Coca-Cola, in a very large brining bucket or food-grade plastic storage bucket (easily found at restaurant supply stores). Make sure the bucket is large enough, as the mixture will foam vigorously. Drop in the ham, adding any rub that has fallen off the meat. Refrigerate and allow to brine overnight. If there's not room in your fridge, brine in a large ice chest with sacks of ice or ice packs.

Remove the ham from the brine, and discard the liquid. Rinse the ham thoroughly and pat dry. Place on a wire rack on a foil-lined roasting pan, cut-side-down for a shank roast (the ankle will taper off into the air, point up).

In a food processor, combine the sage, parsley, 1 tablespoon each of the peppercorns and salt, coriander, and the cloves from half a head of remaining garlic. Turn on the food processor, and dribble in the grapeseed oil until you form a smooth paste. Rub the paste all over the ham, and allow it to sit on the counter at least an hour to come to temperature.

In the meantime, make the glaze by whisking together the sugar, juice, vinegar, and chiles, and bring to a boil. Reduce until thickened, stirring occasionally, about 10 minutes. Once the glaze is made, preheat the oven to 350°F.

Place the roast in the oven and cook it about 2½ hours, or until the internal temperature reaches 145°F. (A remote thermometer will make your life easy here.) After the first hour, baste the ham liberally with glaze every half-hour.

Remove the ham from the oven, and cover loosely with foil. Allow it to rest for at least a half-hour before carving in ¼-inch slices.

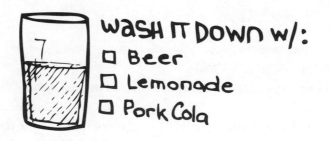

WASH IT DOWN w/:
☐ Beer
☐ Lemonade
☐ Pork Cola

Snowy White Solid Deliciousness for Baking or Frying (Lard)

If you order a whole or half-pork, please trust me and order the fat. Back fat is the thick layer of fat that runs right under the skin along the pig's back. Leaf lard is the fat that surrounds the kidneys. While leaf lard is considered the holy grail for baking, I've happily baked with back fat lard and noticed no meaty odors or flavors. Fat that comes from pastured pigs has an extra bonus: vitamin D and a high level of mono-unsaturated fat (that's the kind that's good for you). Now, I don't think you should dollop it on your breakfast cereal, but it fries up a mean chicken and produces one of the best pie crusts I've ever made. Like the pork itself, you'll find the lard is light-years away from the boxes of hydrogenated manteca in grocery stores. It's a snap to render the fat, and you get a bonus of crispy cracklings that you can sprinkle on your cornbread batter, float on a bowl of Asian noodle soup, or munch with salt.

> Back fat or leaf lard
> Cheesecloth or muslin

Go over the fat carefully and remove any large pieces of remaining skin, flesh, or blood spots. Chop the fat into small, ½-inch cubes, or put the frozen fat through your food grinder to make the rendering process even easier. (If you grind the frozen fat, reserve some of it to mix with ground meat for sausage.)

Pour ¼ cup water into a Dutch oven or other heavy-bottomed pot, and add the chopped or ground fat. Put the pot over the lowest flame your stove can muster, then go about your business. You can stir every once in a while. When you begin to hear sputtering and spitting, it's time to start paying attention. You want to wait just until the solids, bits of skin and meat, fall to the bottom of the pot, then remove the pot from the heat. Pour the fat through a sieve lined with a double thickness of cheesecloth or muslin into a clean container. Immediately pour the lard into sterilized canning jars and screw on the lids. As the lard cools, the lids should seal, preserving the lard a bit longer. Though you can keep it on the counter, I like to keep one jar in the fridge and the rest in the freezer.

A Summer Pie

I don't know about where you live, but in Seattle pie is the new cupcake. Consequently, maybe you make pie all the time. Well, goodie for you. I don't, or didn't, until the prospect of making the ultimate lard/butter crust with my very own rendered lard got my juices flowing. Now, pies fill the house with incredible aromas. You get to make little cinnamon sugar scraps for the kids, who think you're the best parent ever, and if a visitor comes by you can offer them coffee and pie. It's so wholesome, it's practically Midwestern.

To make this crust you'll need a kitchen scale. If you bake a reasonable amount you should have one anyway, so I'm doing you a favor. With this mix of berries I find cinnamon weird, but if you want to add a speck of nutmeg, I won't stop you.

MAKES 1 DOUBLE-CRUST 9-INCH PIE

12½ ounces all-purpose flour

1 teaspoon salt

½ pound unsalted butter, cut into cubes and well chilled

2½ ounces lard, chilled

⅓ cup ice water

In a bowl, whisk together the flour and salt. Add the butter and lard, and cut it into the flour with a pastry blender until the fat is the size of peas. Add just enough ice water to form a dough when the mixture is pressed together. Alternately, you can pulse the flour and salt in a food processor to blend. Add the chilled fat, and pulse until it is covered in flour and the size of peas. Dribble in enough water to form a shaggy mass. Remove the dough from the food processor, and knead a few times to bring the dough together. Divide into two disks and wrap each in plastic. Refrigerate until firm.

To finish the pie:

5 cups mixed raspberries, blackberries, and blueberries

⅔ cup sugar

Pinch of salt

Good squeeze of lemon juice

Zest of 1 lemon
2 tablespoons tapioca
1 tablespoon butter

Preheat the oven to 400°F.

Toss the berries with the sugar, salt, lemon juice, zest, and tapioca. Roll out one disk of the dough and line a pie plate, allowing for a 1-inch overhang. Pour in the berry mixture, and dot with butter.

Roll out the remaining disk to just fit the top of the pie dish. Fold the overhang up and over the edge of the top crust. Go around the edges of the pie, and pinch the dough between your thumbs and forefingers to seal. If you like, brush the top crust with a little water and sprinkle it with a little coarse turbinado or sanding sugar.

Cut a few slashes in the top crust, and bake the pie for 45 minutes, or until the crust is golden and bubbling through the slashes. If the crust browns too quickly, shield the edge of the dough with foil. Allow the pie to cool on a wire rack before serving.

Romeo's Clams with Roasted Juliet Tomatoes and Bacon

Since clams only take five minutes to cook, I feel justified in asking you to slow-roast some tomatoes as part of this recipe. It's idle cooking anyway—just let them linger in a slow oven. For me, the tomatoes preceded the clams: I had a bounty of Juliets in the yard and thought to roast some as a variation on the fresh tomatoes we'd been eating for weeks. (If you don't have Juliets—adorable little egg-shaped beauties—use Romas and cut them in quarters, or cherry tomatoes left whole.) While I was roasting tomatoes and crisping bacon for a strata to put up for the next morning, leave it to Romeo to say that he really, really wanted clams for dinner. He is a wise child. By the way, he'd like me to emphasize that it is necessary to have a good baguette and lightly salted butter to accompany the dish.

SERVES 4

8 Juliet or other "salad tomatoes," or 1 pint cherry tomatoes
Olive oil
Kosher salt and black pepper
6 sprigs thyme
¼ pound bacon ends, cut into small cubes or strips
3 garlic cloves, sliced
2 pounds Manila clams
1 bottle champagne, Prosecco, or dry white wine

Preheat the oven to 250°F.

Wash the tomatoes and dry them. Slice in half lengthwise, and place in a bowl. Toss with olive oil to coat, a liberal sprinkling of kosher salt, pepper, and the leaves stripped from the thyme stems. Line a sheet pan with parchment paper, and arrange the tomatoes on it, cut side down. Roast in the oven for 2 to 3 hours, or until the tomatoes are shriveled and tacky.

Heat a large, heavy-bottomed pot over medium heat. Cook the bacon, stirring, until crisp. Remove it to paper towels to drain, leaving the fat in the pot. Increase the heat to high, add the garlic to the pot, and stir. Add the clams, and nestle the tomato halves among them. Add a splash of champagne (you know what to do with the rest), cover the pot, and cook for about 5 minutes, just until the clams open and release their juices. Sprinkle the bacon on top, and toss. Serve in a large, shallow bowl.

Chinese–Italian–Kosher Pork

Including Marcella Hazan's classic recipe, I have a good handful of formulas for cooking pork in milk, a simple, delicious, if monochromatic Italian braise that depends on garlic and bay-scented whole milk to perfume and gently cook a hunk of pork. I was very excited at the prospect of making this dish with my first pig, only to have my dreams dashed by Romeo's temporary diagnosis of gluten, dairy, and egg intolerance. Yes, aside from meat, the holy trinity was now off-limits. But one day I got an idea. There is a small Chinese tofu factory not far from my office, which sells incredible fresh tofu and soy milk. We routinely had Japanese-style ramen for lunch, made with pork broth unctuous with soy milk. Could I not braise the pork in soy milk instead, using Asian spices? The dish might even end up kosher, as it would no longer combine meat and dairy! . . . I know, I know. My heart was in the right place. And it works—with the soy serving the same creamy purpose as cow's milk, even ending up as the same unseemly curds. Get out a stick blender if it bothers you, but it's so luscious, I don't think it will. A sprinkle of fresh coriander makes it a little less beige.

SERVES 4–6

3–4 pound pork loin roast
2 teaspoons ground coriander
½ teaspoon ground Szechuan peppercorns
Kosher salt and black pepper
Canola oil for browning
1 garlic clove, crushed
2 star anise
1 stick cinnamon
1-inch swath of orange zest, removed with a vegetable peeler
4 cups fresh soy milk (not sweetened)

Dry the roast and rub it carefully and well with the coriander and Szechuan pepper, and kosher salt and black pepper. In a Dutch oven, heat a bit of canola or grapeseed oil on high, and brown the roast well on all sides. Remove the pork and set aside. Add the garlic, star anise, cinnamon, orange zest, and soy milk, and bring to a gentle simmer.

Return the pork to the pot. Partially cover the pot, and simmer for 1 to 1½ hours, or until the roast registers 155°F in the center. Remove the meat, tent it with foil, and allow to rest.

Remove the orange zest, cinnamon, and star anise, then reduce the soy milk over medium-high heat by half. Slice the pork thickly, and spoon the sauce over it.

Spanish-Style Chorizo

Adapted from a loose recipe in Lindy Wildsmith's Cured, these are the sausages I hung in my basement, and also the same sausages that my brothers smoked and froze. We used hog casings during the great Thanksgiving sausageathon, but in the end I decided that lamb casings, with their smaller diameter, were preferable, especially for air curing. If you have a grinder, you don't need to use ground meat: Simply cut the side pork or belly and loin into 1-inch cubes and place them in the freezer until partially frozen. Toss the meat with the seasonings and wine, then put everything through the grinder, twice, to thoroughly mix. Allow the mixture to cure for a day or so before stuffing the casings. If the lack of curing salts gives you pause, consult a charcuterie book for dry-curing instructions and suggested measurements.

MAKES 9 POUNDS

4½ pounds side pork, ground

4½ pounds pork loin, ground

3 tablespoons smoked paprika

1 head of garlic, peeled and minced

1 tablespoon dried thyme, powdered (optional), or 3 tablespoons fresh, leaves only, minced

5 tablespoons coarse sea salt

½ cup dry white wine

Lamb casings (available at certain butcher shops and online)

The day before stuffing, mix the pork and seasonings thoroughly in a non-reactive bowl. Allow the mixture to cure for at least 1 day, and up to 2, in the refrigerator. When ready to stuff, place the lamb casings in fresh, cold water, and separate out the skein.

Stuff the seasoned meat into the casings. Hang the sausages in a dry, cool place for three weeks to cure (I used my basement, with a fan to circulate the air). They may also be smoked, cooked fresh immediately, or frozen and used within 1 month.

On Making Sausage. In Large Quantities. With Your Siblings.

○ ○ ○ ○ ○ ○ ○ ○   ○ ○ ○ ○ ○ ○ ○ ○   ○ ○ ○ ○ ○ ○ ○ ○

You know the old saw about not wanting to watch sausage being made, and perhaps there is no other food that invokes bad-meat-freakout paranoia more than sausage. And there are reasons for that. Conversely, let me tell you what an utter joy it is to take your very own meat, or good meat from an alternate source, season it just so, then turn it into sausages. Lamb sausages, pork sausages, kielbasa, *morcilla*. You might not believe me when I tell you there is nothing standing between you and the sausage of your dreams, but it's nearly close to true.

The day after Thanksgiving, the brothers and I all converged on Dee's house with a total of 50 pounds of meat in hand. Dee has an enormous smoker, as I have mentioned, as well as an industrial-size sausage stuffer. We had bags of seasonings and Prague powder and plastic tubs for mixing. I brought the hog casings I got from my local butcher, packed in salt. (You can also order casings online, both hog and sheep as well as a host of synthetic casings.)

It was cool outside, though not technically freezing, making the stuffing and linking on the outdoor table a more enjoyable endeavor. One person held down the stuffer because we'd lost the clamp that held it to the table, which was now covered completely in plastic wrap. One person cranked. One person pricked the sausages as they were made, to release any air bubbles. One person maneuvered the sausage snake and twisted it into links as it was made. One person monitored Dee's dogs to ensure they didn't eat the sausage. Another looped the sausages and hung them up to dry a bit before they went into the smoker. The nephews mostly hung around making phallic jokes, which I guess was developmentally appropriate.

LAM's sausage maker

When it was done, we had made enough chorizo and pepperoni to keep four families happy for a good while, and it was more fun than watching football. If you have numerous siblings to corral, or a bossypants partner you want to loop into some serious sausage-making, following are a few things to keep in mind no matter what they say.

» Test your seasonings before you stuff. Fry up a patty and taste it for salt and seasonings, and adjust as necessary. Once they're stuffed, it'll be too late.

» Hogs have big intestines; ergo, casings made from hogs make big sausages. Sheep, a bit more diminutive. Especially if you want to cure your sausages, the diameter will affect the curing rate and the time required before you can eat them, as well as the time it takes smoke to penetrate the sausages if you go that route.

» A hand-cranked sausage stuffer is nice and easy to control, but you can also do it with a funnel or an attachment for your mixer. They both work, though the funnel is a *lot* of work and requires two people. I find it especially fun to get a lot of people in on the action, like a barn raising.

» There are a number of sausage schools of thought, from the "let it dry for a while in the salt air" to the "cure at 75 percent relative humidity at 58° until the weight of the sausage (in micrograms) has decreased by 20 percent." Dee falls into the latter category (and the one I suggest for beginners and litigious readers), while I explored the former, just to see how it would turn out. It was a fun adventure, and I successfully air-cured some of my chorizo, producing a more textbook Spanish-style chorizo. But because my humidity varied wildly—I was curing in a basement room, not in a wine fridge or other curing chamber—the texture suffered. But if you really, really want to make salami, and keep making salami, read the scores of books out there, cruise the sausage blogs—of which there are many—and devise a good system that works.

» You will need to decide how to go on the nitrite/nitrate issue. I used a bit of "pink salt" or curing salt in making my chorizo, because I knew I wanted to air-cure it and was concerned about botulism, the result of tangling with *Clostridium botulinum* in low-oxygen environments. However, I attended a charcuterie class taught by people who encourage bacterial growth with buttermilk and wouldn't think of adding curing salts to their products, thus emphasizing the *terroir* in a more old-school way (see Reflections on Butchery in Chapter 4).

And in fact, you don't have to engage with curing salts, or casings, huge plastic-wrapped tables, or any of this craziness! Take ground meat and fat, season it, shape it into patties or wrap it around a skewer and cook it. Do you know what they call that? Sausage. Use it immediately or freeze it for later.

Ma Po Doufu

"Pock-marked mother's bean curd" is how this famous Szechuan dish's name is usually translated, and knowing what the final dish looks like, I get the connection. When Dee gave me his big, bold, peppery rendition, I asked if his wife, Victoria, could verify that ma po *really meant "pock-marked mother." "How the hell would she know?" he replied, and it's true that Vic speaks only enough Cantonese, or maybe it's Mandarin, to order her favorite foods and talk to her granny. In a very un-Chinese move, Victoria sometimes makes this dish with ground turkey. I will pretend not to know that. Serve it with lots of bright vegetable sides, like* gai lan, *to make the health nuts happy.*

SERVES 4

3 squares (4½ ounces each) soft or firm tofu

Oil for frying

6 ounces ground pork or beef

Small bunch green onions, trimmed and minced

6 garlic cloves, peeled and finely chopped

2 teaspoons grated fresh ginger

2 tablespoons fermented black beans

2 pickled hot red chiles, minced

4 tablespoons light soy sauce

2 tablespoons hot bean paste

1½ teaspoons sugar

1 tablespoon Szechuan peppercorns, toasted and ground

1½ cups pork or chicken stock

2 tablespoons cornstarch mixed with enough cold water
    to make a slurry

Cut the bean curd into ½-inch cubes, and place in a sieve to drain.

Heat 2–3 tablespoons of neutral oil in a wok over very high heat. Stir-fry the pork, most of the green onion, and garlic until the meat is half cooked. Add the ginger, black beans, chiles, soy, and hot bean paste, and stir-fry for 1½ minutes. Add the sugar, a pinch of salt (if needed), Szechuan peppercorns, and stock, and bring to a boil. Add the bean curd cubes, and simmer for 5 minutes. Carefully stir in the cornstarch slurry, and bring to a boil to thicken. Garnish with the remaining green onion.

Smoked Pork Chops with Fried Green Tomatoes

Fried green tomatoes are a Southern thing, but they really should be a Pacific Northwest thing. After all, they have sun to ripen their fruit, while we…not so much. There are about as many ways to batter and fry them as there are ways to make clam chowder. You can soufflé them, deep-fry them, coat them in cracker meal, flour, or cornmeal. I like the crunch of cornmeal with little fuss. (My life is already complicated.) Smoked pork chops are a Southern thing, too. They're delicious, and rich, and I'm not one of those people who usually call out foods for excessive richness. The green tomatoes add just the right acidity, like preserved lemon in a tagine, but they're puckery—don't expect your children to gobble them down even though they are fried. It's fun watching them try them, though, like feeding babies pickles just to see them screw up their faces. If you use your green tomatoes to make relish, chutney, or chow-chow, forget about frying them and spoon some of that to the side of the warmed chops instead.

SERVES 4

1 quart apple cider
½ cup apple cider vinegar
3 garlic cloves, crushed
1 cup kosher salt
1 tablespoon black peppercorns, cracked
4 bone-in pork chops, 1 inch thick
4 large, firm, green beefsteak-style tomatoes
½ cup flour
Cayenne pepper (optional)
2 large eggs
Lard or bacon fat for frying
1 cup cornmeal

Combine 1 quart water, the cider, vinegar, garlic, salt, and peppercorns in a non-reactive bowl or tub, and stir well to dissolve the salt. Add the pork chops and cover the bowl. Refrigerate it overnight.

Remove the chops from the brine, and rinse and pat dry. Allow them to come to room temperature while you prepare the smoker. A temperature of around 220–225°F works best, and I use apple wood chips for smoke. At 1 inch thick, the chops should take nearly 4 hours to cook through. If you have thinner chops, check the meat earlier, starting at around 2 hours. They should feel firm to the touch.

Slice the tomatoes thick. Season the flour with plenty of salt and ground pepper, and cayenne for spark, if desired. Lightly beat the eggs in a shallow bowl. Heat a cast-iron skillet on medium-high, with enough lard or bacon fat to amply cover the bottom.

Dredge the tomato slices in the seasoned flour, then dip them in the egg, and then finish them in the cornmeal. Fry just until golden brown. While you are frying the tomatoes, heat a bit of oil in a large pan, and sauté the chops until just warmed through.

8

. . . .

GOAT

My introduction to goat came nearly a decade ago, the day we moved into our house in Southeast Seattle. The neighborhood has as many halal butchers as it does *pho* restaurants, and that's saying something. It was nearly seven o'clock, but it was still hot and sticky, and even if the kitchen hadn't been piled to the rafters with boxes, it was the kind of day when no one wants to cook dinner. Husband Hendel and I drove down the road to a Somali restaurant we'd spied, reckoning any time is a good time for food adventure.

We entered gingerly to find each table occupied by a single Somali man, many of them cab drivers. We approached the counter and studied the menu, not knowing the first thing about Somali food. Ethiopian we knew, West African we knew. Somali, on the other hand, was new. Looking at our faces, the gracious woman behind the counter told us what we should eat, and we dutifully followed her instructions, including ordering a goat dish that she called the national dish of Somalia.

We couldn't wait! It was stifling inside the small restaurant and we got our order to go, eagerly anticipating eating dinner al fresco on our new, so-expansive-it-was-nearly-Southern porch. As we accepted our boxes and bags a tall, thin man approached us. From the looks we had received in general, I would say young white people were not frequent patrons. "Thank you so much for coming here," he said. "I hope you enjoy the food. Thank you, thank you." He bowed slightly and left. The man, the owner explained to us, was a regular patron. As yet unmarried, and with his mother back in

Somalia, he ate dinner in the restaurant nearly every night. The graciousness and acceptance were overwhelming. How much more in love with this neighborhood could we possibly get?

We floated back home on the goodwill of the Somali people and unwrapped our food on the porch. Now, I've had Somali food since, and it's quite good—that wasn't the problem. But as we chewed, both my husband and I realized this was either the goatiest goat that had ever been cooked or simply bad goat, and I don't mean naughty. We looked at one another, grimaced, and spat it out, unable even to chew and swallow.

"Is it bad, or *bad*?" he asked me.

"I don't know," I admitted, swishing out my mouth with beer.

"I'm just happy we got it to go. They were so nice, we would have had to eat it," he said. I had to agree. We chalked it all up to food adventure gone awry and fed the leftovers to the dogs, who nibbled cautiously at the meal. (If you knew my dogs, that's saying something.)

As anyone who has gotten ill off of tequila can tell you, the body is conditioned to respond negatively to foods that have previously produced an ill effect, especially foods with pronounced flavors. Thankfully, we were prudent enough not to reject goat outright (as for the dogs, I don't know), and it was with excitement that I got ready, some time later, for another goat adventure at the farmers market.

When I had inquired about whole goats with the halal butchers in my neighborhood, they couldn't tell me where their goats were raised. But I wanted to talk with the person who was doing the actual goat growing. Because the meat was halal, slaughtered in accordance with Islamic standards, I could infer some important things about it: that it was butchered with one merciful and swift slice across the neck, that it did not did travel great distances to slaughter, and that it was bled out completely before the carcass was butchered. But I wanted to know more. (By the way, Caribbean and Latino markets usually carry goat, too, if you want to try some before ordering a whole one.)

So in my quest for whole animals I decided to try the goat guy at the farmers market. I had noticed him before, because he sold unpasteurized and unhomogenized goat milk for making cheese, something I wanted to try someday. Turns out he also had an ice chest full of frozen, cryovacked goat parts for sale, from rib chops to loin chops to goat burger. I bought some loin chops from the goat man, along with beef and lamb from an adjacent grower I was thinking of ordering from.

This goat was a revelation. Despite my mother's childhood memories of musky goats kept far away from the house because of their smell, I can only describe the meat as sweet and mild. It had a bit of chew to it, to be honest. The boys gnawed away on their chops like dogs on deer antlers, but they agreed the flavor was superb. I was so impressed I was prompted to post "OMG—goat is delicious!" on my Facebook page, wanting to spread the news far and wide. This immediately prompted ridicule from my nephew; evidently I'm too *old* to co-opt the kiddies' texting lingo without embarrassing myself. But for goat, I'd do it again. We returned to the goat guy the next week and I ordered one whole. "Shorthorn" came to live in our freezer shortly thereafter.

What I will tell you is that the Somalis, and most of the rest of the world, are onto something in eating goat meat. Urban meat eaters especially should consider the animal. Goats are a good animal for us—useful, compact, and tasty. On top of that, goat is incredibly lean, low in fat, and high in iron, with the same calories as chicken. Heck, in Seattle we use them to mow our blackberries, and we can even grow them in our backyards, thanks to recent legislation passed by the city council. Maybe more country folk need to give them a go, too. When I first told Cousin Kate that I was going to order a whole goat, she asked what they tasted like. Well, what she really said was, "Are they really gamey? Like cougar?" With a straight face. I can't make this stuff up.

Based on Cousin Kate's testimony, I can say that goat tastes nothing like cougar. Though some people enjoy older goat with a stronger flavor, in the United States we generally don't let kids get old enough to taste musky. They are usually slaughtered before they're nine months old, at maximum a year.

What Kind of Goat Can I Get?

In general there are a lot fewer descriptive terms thrown around when dealing with goat than with other meats. But there are some you will definitely run across in cities, less so outside of major metro areas, especially those with low goat-eating populations.

Organic. Usually this means the goats were raised entirely on forage, without additional feed or without processed feed. This may also mean they were raised without the use of dewormers (worms can be an issue with goats) or antibiotics, or that they only received medications when necessary to restore them to health—ask your grower for specifics. Organically raised goats are free of growth stimulants and hormones. Goats may also be fed hay, silage, and crop residue without grain, as well as minerals and vitamins. "Grass-fed" is not often seen in relation to goats, though "foraged" can be.

goat in your
underwear drawer

Should I Care What My Goat Eats?

When I first asked what food my goat ate, the grower's answer was pretty much "everything." And while goats are reputed to eat everything under the sun, including tin cans, in general you can assume your goat ate available scrub, bushes, and trees, perhaps in addition to minerals and grain as fed by their owners. I know for a fact that Shorthorn ate a lot of blackberries and dandelions.

What Breed of Goat Should I Choose?

Unlike with cows, goat meat is regularly produced from many breeds, including both meat and dairy goats. Dairy goats are less hardy and kid less often than meat goats. But since they're bred to produce milk, let's just say that little boy goats aren't particularly useful out there on the farm, and some growers will keep them as meat goats if it pencils out for them.

The LaMancha is a tiny-eared breed of European stock, while Nubians have distinctive "Roman" noses and long, dangling ears. Some of the more common dairy breeds include Alpine, Oberhasli, Saanen, and Toggenburg, large goats with upright ears; you can imagine the bell around their necks. Alpine goats tend to have bigger kids at birth, so if you're buying an "Easter goat" or a young spring kid, this is a fine option.

Several breeds are raised specifically for meat production. These are more deeply muscled and hardier goats, often allowed to run wild over scrubby pasture that would never support cattle. For you, this means goats that are available nearly year-round and carcasses with larger legs and more meat. These breeds include the Spanish Meat Goat, often just called Spanish or Black Spanish; the white Kiko, imported from New Zealand, which is a shade larger than the Spanish and has Nubian and Saanen blood; and the currently popular Boer goat from South Africa, a long-eared white goat with a distinctive red head. Boers are known as "double-muscled" because they can build so much muscle in less time.

By far my favorite, simply for the romance of it (although the goats I've purchased have been Boer crosses), is the Tennessee Meat Goat, all specimens of which are descended from a myotonic flock from the nineteenth century (*myotonia* is defined as "the inability to relax voluntary muscle after vigorous effort"). Have you ever heard of fainting or stumbling goats? These are afflicted with a physical condition that causes their muscles to temporarily stiffen and lock up, especially when the goats are startled or excited, to the point that they can fall over sideways. The condition doesn't harm them, and the spells last only about ten seconds. Reputedly, it has the effect of producing especially heavy muscles in the rear legs and more tender meat. Even if you don't end up ordering one for consumption, I highly recommend watching YouTube videos of myotonic goats in action. They're even better than Ninja Cats.

What Cuts of Goat Can I Order?

One of the great things about ordering a small animal is that you can eat it up, see what your favorites are, and choose different cuts the next time. Unlike with other animals, I love having plentiful ground goat and lamb. I have a grinder attachment for my KitchenAid so I can grind it myself (a task the boys are terribly fond of), but it's easier to take a package out of the freezer.

In general, I find goat and lamb nearly interchangeable in most recipes, though there are some bold, spicy dishes—like Trinidadian goat curry or Mexican *birria*—in which I prefer the stronger flavor of goat to that of any other meat. And without exception, goat is tougher; the flavor is nice, but expect to use your molars unless you cook it long and slow. With this in mind, I don't feel as covetous of goat leg as I do a bone-in lamb leg, a cut I prefer to roast as simply as possible. You absolutely can give it the same treatment, but don't expect tender. Because goat can be tough, it's not a terrible idea to let the meat rest in the fridge unwrapped for a bit before cooking, "aging" it for a day or two.

Primal Needs

I have not yet encountered a separate cut sheet for goat. What I have found instead are growers who tell you to find a cut sheet for lamb on the Internet and make the same choices, or who make recommendations for cuts to choose from either a lamb or a goat. A goat is divided into the same five general primals as a lamb, and if you have general knowledge about lamb cuts, you will have a head start in ordering your goat cuts. Think lamb, only smaller.

Though some don't think of it as a primal and lump it in with Shoulder, I have to start off by singling out **Neck** here, especially on a goat. If you order neck slices, and I think you should, you'll be rewarded with lovely meat that takes to braising like a bloodhound to a scent. If you don't specify that you want neck slices, they will probably be ground. Which isn't a terrible thing, but still.

Goat Neck Cuts:
Ground meat
Slices (bone-in)

We move on then to the **Shoulder**, a flavorful primal full of all sorts of fun stuff, from the shoulder blade down to the front shank. This meat can be bone-in or boneless, rolled and tied. You can cut stew meat from the shoulder or grind it. You can even cut shoulder blade chops from this primal, though don't expect them to be as tender as loin or rib chops, especially with a goat. I have to say that as much as I love lamb shanks, I love goat shanks even better.

Goat Shoulder Cuts:

Stew meat   Blade chop
Ground meat   Shank
Shoulder or blade roast

Goat **Breast** is part of the forequarter and contains some ribs. It is oblong and lean, with fat usually covering one of the sides. (Perhaps you've heard of breast of veal, a more common cut than breast of lamb and certainly breast of goat.) If you want the breast for roasting or braising, you will need to ask; this is never a cut that a regular processor will give you. You will receive goat riblets that are fine for cooking if you remove the "fell"—a papery membrane that's not fun to eat—before doing so. These riblets have thin layers of meat and fat and need braising or moist heat.

Goat Breast Cuts:

Breast
Riblets
Ground meat

The **Rack** of goat is the top back part of the animal, and rack of goat is as good as rack of lamb, if not as tender. A goat rib roast includes rib bones, backbone, and the ribeye muscle. Rib chops have backbone and can contain rib bone. Cook them fast to keep them tender.

Goat Rack Cuts:

Rack
Rib chop
Rib roast

The equally tender **Loin** falls behind the rack and breast. On a goat, loin chops are tiny, glorious little things that cook in an instant under a broiler, on a grill, or in a pan. The portion size is lovely for those wanting to accent a meal with meat rather than make a hunk of it the centerpiece. Loin chops contain part of the backbone, the eye of loin muscle, and flank, while double chops contain the top loin and tenderloin and no flank. The tenderloin could be asked for as a separate cut, but it's just the tiniest thing on a goat.

Goat Loin Cuts:

Boneless loin roast	Double loin chop
Loin chop	Tenderloin

Ah, the **Leg**—somehow elegant and medieval at the same time. To me, a leg of goat feels somehow more bad-ass than a leg of lamb, and maybe more biblical. Tougher, too, as I might have mentioned. Asking for a whole boneless leg that you can butterfly or cut up and marinate before cooking isn't the worst idea. A whole leg contains both the sirloin (the section with hipbone) and the shank (with round bone). You can also order sirloin chops or have the leg cut into cubes, or ground. The rear shanks are meaty and glorious.

Goat Leg Cuts

Whole leg, bone-in or boneless	Sirloin roast
Sirloin chop	Lamb shank

Goat **Innards** aren't as strongly flavored or as big as the innards of other animals, therefore I don't mind them as much. Especially if you have a grinder. I will tell you that a little cryovacked goat heart looks exactly like what it is, and it's fairly precious.

How Much Will My Goat Cost and How Much Will I Get?

Maybe because you are dealing with a smaller animal, buying goats is usually less complicated. You pay one price to the producer, which includes the cut-and-wrap and the "harvest" or kill fee. Shorthorn was itemized as costing us $6.99 per pound hanging weight, plus an additional $1.50 per pound for the cut-and-wrap. For a 34-pound goat, hanging weight, this came out to about $289.

I've also found goats sold for set prices by the half and whole. The grower should give you a range of either weights or take-home meat in pounds that you will get for your price.

Questions to Ask Your Grower:

» Are half-goats available?
» How much deposit is required?
» At what age are your goats slaughtered?
» What is the average hanging weight of your animals?
» Is the slaughter and processing included in the fee, or will I pay the processor separately?
» What time of year are your goats available?
» Where can I pick up my meat?

How Much Meat Will I Get?

This is the glory of goat! They're little enough that a six-year-old can wheel a whole one, cut and wrapped, around a farmers market, as Romeo can attest. A whole goat will easily fit in your regular freezer, and individual cuts defrost quickly and are nicely sized for singles or couples.

The dressing percentage of a goat is affected by all sorts of factors, including how hairy the goat is, the gender, and how long it was off feed or water before slaughter. About 50 percent is a reasonable average, though you should ask your grower what is true for his or her goats. This means that a 60-pound goat will yield a carcass of about 30 pounds. The amount of meat you get from the animal is about 36 to 41 percent of the live weight, so a 60-pound goat will yield anywhere from 22 to 25 pounds of meat.

The exact size depends on the age of the goat you buy as well as the breed. Shorthorn was a Boer cross that came in at 34 pounds hanging weight, and he was a little bitty thing. Depending on the age of your goat, you could receive a heck of a lot more. Make sure you ask for the live weight at which your goat is slaughtered. Some folks prefer the taste of a more mature animal, and if your grower caters to that clientele you might have a distinctly different experience and a lot more meat on hand as well.

How Much Freezer Space Do I Need?

If you're ordering an average goat weighing 60–100 pounds on the hoof, the freezer compartment of an average refrigerator/freezer unit will hold your goat with room to spare.

Who Will Kill My Goat and How?

Your goat will either be stunned or shot and then have its throat cut, or simply have its throat cut with a very sharp knife. Because goats are on the small side and relatively tame, many small producers kill and process their own carcasses in order to save money.

When Will It Be Ready?

In my case, from order to pickup took about a month, as my grower only killed kids in batches. (That just doesn't sound right, does it?) He waited for enough orders to come in to have them processed. If you work with a larger grower, the animal may be ready sooner than that. Once it is slaughtered, the meat is available fairly quickly, as nothing needs to be processed or aged further.

RECIPES FOR GOAT

Qorma Pilau

This Afghani rice and goat dish is both slightly exotic and totally mild and accessible. I love the way it transforms some pretty humble ingredients into Sunday night dinner. The riblets make it quite rich; substitute leaner goat or lamb cuts for a lighter flavor.

SERVES 6

3 cups basmati rice
½ cup olive oil
3 medium yellow onions, chopped
2 pounds (approx.) goat riblets, neck, or other goat on the bone,
   cut into large chunks
½ cup yellow split peas
1 teaspoon each cumin seeds, cardamom seeds, cinnamon, and
   coriander seeds, ground together
1 teaspoon saffron, soaked in warm water
Fresh cilantro (optional)

Rinse the rice well, then soak it for an hour.

Heat half the olive oil in a large Dutch oven on high, and add the onions. Sprinkle the onions with salt and half of the spice mixture, and fry the onions until soft and brown. Remove the onions and set them aside.

Add the remaining olive oil over high heat. Add the meat, in batches if necessary, and brown on all sides. Return the onions to the pot, add the split peas, and add enough water to cover the meat halfway. Cover the pot, and cook until the peas and meat are quite tender. Stir in the spice mixture.

Preheat the oven to 300°F.

Drain the rice, parboil in briskly boiling salted water for 3 minutes, and drain. Spread half of the rice over the bottom of a large casserole or Dutch oven. Scoop out about 1 cup of the juices from the other pot, and mix with the saffron, allowing the saffron to bloom, or dissolve in the liquid. Using a slotted spoon, add the meat, onions, and split peas, and top them with the remaining rice.

Pour the saffroned meat juices over the top of the rice and meat mixture, (use the rest of the juices for soup, if you have a lot). Cover the pot tightly, and bake for 45 minutes to 1 hour. The rice should be pleasingly plump and the whole thing fragrant and glorious. Spoon it all out onto a broad platter. I like to garnish mine with chopped cilantro, but you don't have to.

Trinidadian Goat Curry

This is one of the very best things you can do with goat neck; I think I could eat it every week for the next five years and never grow tired of it. I made this awfully fiery the first go-round, but less so on my later bouts because the children loved it, too, especially with the special flatbread that accompanies it (see the Roti recipe after this). I use all the neck and a shoulder, but you can make it with any cuts, adjusting your cooking time accordingly: Loin or leg, for example, don't need as long to stew. It's also fabulous with lamb, or mix it up if you happen to have both in your freezer. Trinidadian curry powder is a little different from what you might be used to. You can make your own or buy it, but make sure there's no cumin in it and that it's very fresh— it makes all the difference in the vibrancy of the dish. This recipe makes quite a bit, but you'll enjoy the leftovers.

SERVES 6

3–4 pounds (approx.) goat meat with bones, such as neck,
 shoulder, or a mix of the two, or use boneless leg, cut into chunks
Juice of 2 limes
2 tablespoons kosher salt
1 tablespoon cracked black pepper
2 habañero chiles, or to taste, stemmed
¼ cup Trinidadian or Jamaican curry powder, or any blend with little or
 no cumin
1 tablespoon garam masala
1 teaspoon ground allspice

Leaves from 4 sprigs fresh thyme

Small bunch green onions, trimmed

4 garlic cloves

Handful cilantro leaves and stems, rinsed

⅓ cup grapeseed or canola oil, plus more for frying

2 small onions, thickly sliced

1 large sweet potato, peeled and cubed

4–5 tomatoes, diced, or ½ 28-oz. can plum tomatoes,
    drained and crushed

1 can coconut milk

The night before you plan to make the curry, rub the goat with the lime juice, and season it with the salt and pepper. In a food processor, combine the habañeros, curry powder, garam masala, allspice, thyme, green onions, garlic, cilantro, and ⅓ cup oil. Purée to a thick paste. Place the goat in a large resealable bag, and pour the paste over it. Seal the bag and squish the paste around to ensure the meat is evenly coated. Refrigerate the goat overnight, turning the bag and squishing it every now and again.

The next day, take the goat from the bag, removing as much marinade from the goat as possible. Reserve all the marinade. Heat a couple of tablespoons of grapeseed or canola oil in a heavy pot or Dutch oven over high heat. Add the goat, and sear on all sides. Add the onions and sweet potato, and cook, stirring, until the onions are limp, about 5 minutes. Add the reserved marinade, tomatoes, and coconut milk, along with enough water to fully cover the meat. Taste for salt. Bring to a boil, then lower to a simmer and cook for 2 to 3 hours, or until the meat is tender. (You could do this in a slow cooker, I imagine. It would also do nicely, covered, in a slow—250°F—oven.)

When the meat is tender, remove the cover and simmer the stew on the stovetop until the curry is thickened. You really can't overcook the curry if you are using neck or shoulder or something with a lot of connective tissue. The meat will simply melt into the sauce, making a meaty, fabulous sludge. Leg will take less time, so watch it carefully. The older your goat, the longer you'll want this to cook. Serve with roti.

Roti

The Seattle Malaysian restaurant Malay Satay Hut turns out incredible food span-ning the whole gamut of influences on Malay cooking, including Nonya, Chinese, and Indian. Their roti canai, a flaky flatbread that you dip into a simple curry sauce, is to die for, and for years I've wanted to replicate it at home. Husband Hendel agreed with that idea, and spent the good part of a year buying me gadgets like electric flatbread griddles and flatbread cookbooks in order to entice me to try. It took Shorthorn the goat, and another island nation with strong Indian influences, for me to seal the deal. Needing an authentic accompaniment to Trinidadian Goat Curry, I sleuthed out a cousin of roti canai that the Trinidadians call "buss-up shut" roti. The name is slang for "busted-up shirt," referring to the fact that the flatbread is cooked on a griddle, then bashed violently with wooden spoons to create a ruffled pile of goodness. It's not as rich as roti canai, but it involves a sporting kind of kitchen violence that can lift you out of a bad mood. And it's the perfect complement to goat curry. Any leftover roti can be frozen and microwaved for 10–15 seconds with fairly good results. This one's for you, mi amor.

MAKES 6–8 FLATBREADS

1 stick butter
4 cups all-purpose flour
2½ tablespoons baking powder
½ teaspoon salt

Begin by melting the butter in a small saucepan over low heat. When it foams, skim the top. Continue to cook until the butter turns golden and the milk solids settle to the bottom. Pour off the clarified butter, reserving and discarding the solids, and let it cool.

In a large bowl, sift together the flour, baking powder, and salt. Add enough water to form a rough ball, starting with about 1½ cups and adding more by tablespoons as needed (up to about 2 cups) to make the dough come together. Add two tablespoons of the clarified butter, and knead the dough inside the bowl to incorporate. Alternatively, sift the dry ingredients into a food processor and run, dribbling in water, just until a ball forms. Then add the butter, and pulse to combine.

Turn out the dough onto a lightly floured board, and knead just until smooth, about 6–8 minutes. Place it in a clean bowl, cover it with a damp towel, and allow it to rest for 30 minutes.

Divide the dough into 6 to 8 pieces, depending on how large you want your roti. (I find dividing into 8 is more manageable for my griddle, but you can make fewer and larger breads if you like.) Form each piece into a smooth ball. Cover the pieces, and let them rest another 15 minutes.

Have ready the remaining clarified butter and a pastry brush. Working with one ball at a time, roll the dough into a circle about 12 inches in diameter. Brush the exposed surface lightly with butter. Take a sharp knife and, beginning from the center of the circle, cut through the dough from the center to the outer edge, as if you were tracing the radius of the circle. Take one entire cut edge and start rolling it away from the other cut edge, clockwise, like a second hand going around a clock, until you have rolled up half the circle. Repeat in the opposite direction with the remaining cut edge, until you meet the rolled-up side. Pinch the two together to seal. Don't worry if your dough has an odd shape, or if the rolled-up sides seem uneven. Your goal here and in the next step is simply to create layers in the dough that will separate when exposed to heat, resulting in a flakier roti.

Stretch the dough slightly, and begin rolling it back and forth along your board with the flat of your hands to even out the thickness, rolling the dough as you would make a snake with Play-Doh. Your snake should be about 6 inches long. Starting at one end, coil the dough in a spiral. Pinch the edge of the spiral, then poke the end into the dough a bit to seal. If it looks irregular, pay no attention—you're going to roll out these balls again, and then bash them with spoons, after all.

Cover the coiled balls tightly with plastic wrap. Allow to rest for at least an hour on the counter, to give the dough a chance to relax.

When you are ready to cook, heat a large griddle, cast-iron or otherwise, over high heat (a large nonstick frying pan will work too. My griddle is the rectangular kind that fits over two burners, so I roll my roti into ovals instead of circles.) Working with one ball at a time, roll out to about ⅛-inch thickness, in ovals or rounds as you desire. Brush the surface of the griddle with clarified butter, and place the roti on top. Cook for about 45 to 60 seconds, or until the bottom side is lightly golden in spots. Brush the top of the roti with butter, then flip it and cook until that side is golden as well.

Now for the fun part! Take two wooden spatulas and just have at the thing, folding and crushing it until it looks like a crumpled shirt. Wrap it in a tea towel to keep warm, and repeat with the remaining roti.

Birria de Cabrito

If I had a Mexican abuelita, I would like to think she would make this version of birria for all important family gatherings. This goat soup/stew works well with all types of meat, from pork to lamb, though goat is by far my favorite. The meat is rubbed with a brick-colored, fragrant paste of spices, chiles, garlic, chocolate, and vinegar and left to sit overnight. The next day, it is steamed for hours, producing a spicy, rich broth that gets ladled around piles of meat in individual bowls. This broth is too spicy for all but the most adventurous of children, though the meat alone should be fine. Using only anchos would lower the heat but would also make me sad, because the different chiles work so well together. I'd rather make the kids extra guacamole and keep the birria all to myself. Cabbage, cilantro, and onion make nice garnishes, and a dash of Fiery Accent Sauce offsets the rich broth perfectly. Serve with fresh corn tortillas and cold beer.

SERVES 6–8

6 guajillo chiles

4 ancho chiles

½ teaspoon cumin seeds

4 whole cloves

1 tablespoon black peppercorns

2 cinnamon sticks

1 teaspoon Mexican oregano

6 garlic cloves

½ cup apple cider vinegar

½ tablet Mexican chocolate

2 tablespoons kosher salt

6 pounds assorted goat parts (or a mix of goat and lamb),
    such as shank, shoulder, and ribs

3 medium tomatoes

½ medium white onion, finely diced

½ bunch cilantro, rinsed, dried, and chopped

¼ head cabbage, shredded

1 lime, cut into wedges

Fresh corn tortillas

Fiery Accent Sauce (recipe follows), for serving

The night before serving, prepare the seasoning paste. Stem the dried chiles and shake out the seeds. Place the chiles in a medium saucepan, and cover with water. Bring to a boil, and boil for 5 minutes. Cover the saucepan and remove from the heat. Allow it sit for 5–10 minutes, or until the chiles are pliable.

Place the cumin seeds, cloves, peppercorns, cinnamon sticks, and oregano in a small mortar or electric spice grinder, and grind them to a powder.

Transfer the chiles to a blender, reserving the liquid. Add the garlic, vinegar, chocolate, salt, and ground spices. Blend on high until you have a thick paste. Add chile-soaking water as necessary to loosen the mixture, but try to keep it thick enough to really coat the meat.

Smear the paste evenly over the meat, turning to coat well, and place the coated meat in a large bowl or two resealable plastic bags. Cover and refrigerate overnight.

On the day you want to cook, place a rack in a large metal roasting pan (I use the V-rack used for large roasts or turkeys), and pour about 1 quart of water in the bottom of the pan. Arrange the meat in a single layer, if possible, on the rack. Cover the meat and rack completely with heavy-duty foil, and crimp the foil tightly around the edges of the pan. (If you need to overlap more than one sheet of foil, you can place a dishtowel over the seam to help keep in the steam.) Set the roasting pan to straddle two burners, over medium-high heat. Once the water reaches a boil, adjust the heat to maintain the steam but to keep the liquid from boiling furiously. Alternatively, you can place the foil-wrapped pan (minus the dishtowel) in an oven at 325°F. Steam the meat for about 4 hours, or until it is very tender and falling off the bone. The juices and melted fat will have mingled with the water, creating a broth in the bottom of the pan.

Transfer the meat to a shallow dish, and cover with foil to keep warm. Skim the fat off the broth and taste for salt. If the broth is too rich, add extra water to lighten the body.

Pour 2 cups of the broth into a blender. Add the tomatoes, cut into large chunks. Purée the liquid with the tomatoes, then whisk the mixture back into the remaining broth. Strain the broth into a saucepan, and heat it until it is steaming hot. In the meantime, remove meat from the bones.

Place about ½ cup of meat in each bowl, and sprinkle with salt. Ladle 1 cup of broth around the meat, and sprinkle it with onion, cilantro, and cabbage. Garnish with lime wedges, and pass tortillas and Fiery Accent Sauce or other hot sauce at the table.

Fiery Accent Sauce

MAKES A SCANT ⅔ CUP

25 chiles de arbol
½ cup apple cider vinegar or white wine vinegar
2 tablespoons lime juice
Scant ½ teaspoon dried oregano
2 large garlic cloves, peeled and crushed
6 black peppercorns, crushed

Place the chiles in a small saucepan, and cover them with water. Bring the water to a boil, then reduce the heat and simmer for 10 minutes. Cover the pan, remove from the heat, and allow to sit for 10 minutes.

Drain the chiles, and place them and the remaining ingredients in a blender. Blend on high until completely liquefied. Add salt to taste, and pulse to combine. Strain the sauce through a fine-mesh sieve into a non-reactive bowl, and set aside to let the flavors meld.

Intolerant Goat Balls

Until Romeo started getting chronic stomach aches, I didn't know how lucky I was. I had never been that mother who has to interrogate other parents about every ingredient in the birthday cake, trying to avoid pain, itching, or death. Romeo had never been that kid, either, and I don't know whose eyes grew bigger, his or mine, when the naturopath said Mo was "reactive" to gluten, dairy, and eggs. Holy cow. This was uncharted territory for me as a cook, and limiting for poor Romeo as an eater. But I rose to the challenge. I made cookies with xanthan gum and used flax seed instead of eggs in our quinoa pancakes. At least, he could have unlimited vegetables and fruit—a good thing, since  *if he were reactive to mango he would have rioted. But there were still so many foods suddenly off-limits, and he was still a six-year-old kid. Luckily, goat was on the permitted list. Since Shorthorn had just landed in our freezer, we pulled out some ground goat and set about making meatballs that were gluten-, dairy-, and egg-free. We made them together, and ate them with homemade tomato sauce and brown rice pasta, and ended up with a dish that didn't feel deficient, just delicious.*

SERVES 4

- 1 tablespoon ground flax seed
- 3 tablespoons unsweetened, unflavored almond or soy milk
- 3 thin slices stale gluten-free bread (if you've eaten gluten-free bread you already know making it stale isn't difficult)
- 1 juicy garlic clove, minced (I would add more but Mo prefers less— you choose)
- 2 passes of a nutmeg across a grater, or the smallest pinch ground nutmeg
- 3 tablespoons chopped parsley
- 1 teaspoon chopped fresh oregano
- 1 pound ground goat (or pork, or beef, or a combination)
- 3 tablespoons olive oil, plus more for frying

Whisk the flax seed into the almond or soy milk, and set aside to thicken. Place the bread, garlic, nutmeg, herbs, and salt and freshly ground black pepper to taste in a food processor, and pulse to form crumbs.

Combine the goat meat with the oil and the flax mixture in a large bowl, and knead to combine. Add the crumb mixture, and mix just until combined. Add up to about 2 tablespoons of water in dribbles if the mixture seems too dry.

Form the mixture into 1-inch balls. Place a good slosh of olive oil in a large sauté pan, and heat over medium-high. When the oil shimmers, add the meatballs in a single layer. You want enough room in the pan for them to roll around a bit; if your pan is small, brown them in batches. Cook, turning as needed, over medium for 8–10 minutes, or until just cooked through.

Roast Chicken with Rice and Goat Stuffing

One Sunday afternoon Atticus and I were in the kitchen ready to cook. We had just picked up Shorthorn the goat that week, and I was excited to try something new with the meat. We also had a kosher chicken in the fridge. Atticus wanted to incorporate the plump pomegranates sitting on the counter. Beginning with a lamb-stuffed chicken recipe from Paula Wolfert, we created a dish that made us both happy. I'm a frequent roaster of chickens, and I like them simple: salt, lots of salt, preferably the night before, an herb sprig or two, a blast of high heat to blister the skin. But stuffing a good bird with rice, nuts, and ground meat makes a roast chicken into something else entirely. At first the initial steaming seemed fussy, but when we ate it, it was brilliant. And Atty's flourish of pomegranate seeds really did make the dish. This is chicken fit for company, though your family will feel very special if you make it for them. Even more so if you make it with them.

SERVES 4–6

Whole chicken, 3–4 pounds, rinsed and dried
Kosher salt
1 pomegranate
2½ tablespoons butter
¾ pound ground goat
⅓ cup pine nuts
2 teaspoons Lam's Syrian Spice Blend (recipe follows)
1 cup brown basmati rice, soaked for 15 minutes and drained
⅓ cup pistachios, coarsely chopped

½ lemon

Cracked black pepper

1 cinnamon stick

1 strip orange peel removed with a vegetable peeler

Rub the chicken with kosher salt, inside and out, and let it come to temperature on the kitchen counter.

Cut the pomegranate in half. Submerge one half in a bowl of cool water, and gently pry out the seeds, keeping the fruit under the surface. The water keeps any errant juice and seeds from flying out onto the counter and staining it and your hands. Drain and set aside. Reserve the remaining pomegranate half for another use.

In a large sauté pan with a lid, heat 1 tablespoon of the butter over medium heat. Add the goat and a pinch of salt, and cook, stirring frequently, until the meat is lightly browned (lower the heat if the butter begins to brown). Add another tablespoon of butter, and when it begins to melt, add the pine nuts. Cook the nuts until lightly toasted. Add another pinch of salt, 1 teaspoon of the spice blend, the rice, and the pistachios, and stir to combine.

Add 2 cups of water and bring to a boil, then cover the pan and lower the heat to a simmer. Cook until the water is absorbed, about 15–20 minutes. Meanwhile, zest the half lemon.

When the liquid is absorbed, adjust for salt and pepper and stir in the lemon zest. Remove the mixture to a bowl or sheet pan to cool.

When the meat and rice mixture has cooled, rub the chicken with the remaining teaspoon of spice mix, along with plenty of cracked black pepper. Stuff the chicken loosely with the meat and rice mixture, then tie its legs together with a bit of kitchen twine. (I'm not usually a trusser, but you've got to keep that stuff inside.) Any extra stuffing can be placed in a ramekin and covered.

Put the chicken in a Dutch oven with a lid. Add 3 cups of water, the cinnamon stick, and the orange peel. Dampen a long sheet of parchment, crumple it loosely, and tuck it around and on top of the chicken as if it were tissue paper in a gift box. Cover the pot and bring the water to a simmer. Simmer for 1 hour. At the 45-minute mark, preheat your oven to 450°F, and remove the orange peel.

Carefully transfer the chicken to a roasting pan with sides. (I find a cake spatula useful for doing this.) Pour the remaining liquid around the chicken. Melt the remaining ½ tablespoon butter, paint the top of the chicken with it, and place the chicken in the oven. If you have a ramekin of leftover stuffing, place it in the oven, too. Roast until the bird is well browned and the drumsticks roll around in their sockets, about 20 minutes. Remove the chicken from the oven and set it aside to rest, tented with foil.

Remove the fat from the pan liquid and reduce the liquid by half. Heat a large platter, and spoon rice stuffing over it to cover. Carve the chicken and arrange the pieces on top of the rice. Drizzle the chicken and rice with the pan juices, or you can pass the juices separately for people to apply as they wish. Sprinkle everything lavishly with pomegranate seeds, and serve.

Lam's Syrian Spice Blend

This nontraditional blend replaces cumin with a touch of cardamom for a Swedish twist, and uses pickling spices for a kick of clove, cinnamon, and allspice. Do your best to find fresh pickling spices for this—you will absolutely taste and smell the difference. Use this for the preceding chicken recipe, or the Lamb Kibbeh recipe, or mix in some olive oil and rub a leg or chop with it. It's addictive, and just spicy enough for interest without risking protests from the minions.

MAKES ABOUT 4 TABLESPOONS

3 tablespoons pickling spices, including a red pepper pod or two
1 teaspoon black peppercorns
1 teaspoon coriander
Seeds from 3 green cardamom pods
1 teaspoon kosher salt

All you do with this is grind the ingredients into a fine powder and store the mixture in an airtight glass jar. You could do it with a mortar and pestle, but, really, do you have the time? I use a coffee grinder—one that I reserve for spices so my coffee doesn't taste like pickles—and I highly recommend it.

Goat Loin Chops with Chimichurri

This was the first dish I made with Shorthorn. If you want to do it up in true Argentine style, then grill up a mix of chops and steaks and sausages, and serve them all with the sauce. This is only one suggestion for what to serve with them. They would also be nice marinated in lemon and olive oil and oregano for half an hour and then grilled. They would be lovely served with a dollop of tapenade. You already have the great meat—the hard part's already done.

SERVES 2

1 cup (packed) fresh Italian parsley

⅓ cup (packed) fresh cilantro

3 garlic cloves

½ teaspoon smoked paprika

½ teaspoon ground cumin

1 teaspoon salt

⅓ cup sherry vinegar

½ cup olive oil

4 goat loin chops

In a food processor, pulse the parsley, cilantro, garlic, paprika, cumin, and salt until chunky. Add the vinegar, and pulse twice more. With the motor running add the oil in a stream, trying not to overprocess. Allow the sauce to sit for at least 10 minutes to blend the flavors.

Heat a grill to high. Season the chops with salt and pepper to taste. Grill over direct heat for 3 minutes per side for medium rare. Serve with chimichurri.

9

• • • • • • • • • • • • • • • • • • • •

LAMB

I always wonder about people who say they "don't eat lamb." Midwesterners who are cautious of goat I will tolerate, even though goat *is* the most widely eaten meat in the world. But for me, people who say they don't eat lamb are as bewildering as those who "just don't like fish" or find cooked onions in a dish "too spicy." I know I've gone on and on about how we all need to come together and get along, but, really, I can't pretend to understand it.

For all its rating a ten for cuteness factor, lamb has a reputation as a Strongly Flavored Meat, so much so that the lamb seems to grow exponentially in my imagination until the little thing could battle King Kong. I will agree with you that grain-fed and grass-fed lamb do have different flavor profiles, with grain-fed animals having much milder meat. But I don't buy lamb to have it taste like chicken, and I prefer the stronger, slightly gamier flavors that grass-fed lambs produce.

While I do love to nibble a lamb chop and find a rack of lamb quite elegant, I like to cook my lamb Middle Eastern style. I like it braised, I like it stewed, I like it ground, and I like it cubed. I love it with cinnamon and turmeric and coriander. Yogurt and lamb are a love match. Tagines are as easy as pie. Make that easier.

Whether you splash it with chimichurri or stuff it with tapenade, I love the big, bold flavors that lamb can not only stand up to but revel in. And yet, as much as I love lamb, I confess to rarely buying it in the store. Why? I don't know. But I find that lamb is my go-to meat now that Puffy Legs is in the freezer. And the whole family is happy he has come to stay.

Lambs are a wonderful choice for urbanists. They're small, for one thing, and not so much of a commitment for those just coming to or experimenting with the whole-animal lifestyle. They are more expensive than goat, but it's much easier to find one. One of the only things to keep in mind if you decide lamb's the one is that (though this is not true for all breeds) lambs are generally born in the early spring—you know, Easter lamb and all that. And you will find tiny little ones around late April. But your animal probably won't reach market weight until fall.

What Kind of Lamb Can I Get?

As with beef, the buzzword for lambs these days seems to be "grass-fed," though you will probably run into many other of the same descriptors for beef as well. Some growers even have variations depending on the time of year. They might be able to raise spring-born lambs exclusively on pasture, but find they need to supplement the diets of those born in other seasons. As with any meat, ask your grower specifically how they are using terms like "organic" or "natural," unless they are part of a certification system.

Grass-fed and finished. Your lamb could be marketed as grass-fed, but finished on corn, wheat, barley, or soy, resulting in a different nutrient profile and milder taste. If the health benefits and taste of grass-fed lamb are what you want, make sure you ask your grower if your lamb was finished on grass as well. If so, it will probably be more expensive because of the extended time it takes to reach slaughter weight.

Pastured. Pastured lambs are iconic animals, little woolly beasts gamboling about in open pastures and not kept in pens or other housing. However, they still might eat by-product feeds—cheap, caloric, and proteinaceous feeds sometimes used in winter months to supplement mature pasture or hay. If you aren't eating a spring-born lamb, it may have received this feed. Ask your grower for specifics.

Natural. Sometimes "natural" means a lamb was grass-fed, sometimes grass-fed and -finished. Other times, it means the lamb was pastured but its diet supplemented with grain, still other times that the grain was not grown with chemicals. You'll need to ask.

Halal and kosher. Although goats are the most common animal raised and slaughtered under the halal practices, you may find lamb, too, though I was unable to find growers who could guarantee certified halal or kosher lambs. In addition to the specific slaughter methods we discussed, certified halal and kosher animals (if you can find them) are raised without the use of antibiotics or hormones and are fed natural feed. Humane growing practices are emphasized.

Should I Care What My Lamb Eats?

Grass-fed lamb, like other grass-fed and -finished meats, is higher in omega-3 fatty acids, as well as vitamin E, beta-carotene, thiamin, riboflavin, calcium, magnesium, and potassium. It also has a more distinctive, deeper flavor and is leaner. For this reason, many consumers have a definite preference for grass- or grain-fed. Try before you buy, if you can.

What Breed of Lamb Should I Choose?

Perhaps more so than with any other animal, I found that whole lambs for sale rarely have their breed mentioned. Instead, they are usually characterized by the time of year they are born (usually the spring) and by what they eat (in part dictated by when they were born). Growers may raise crossbreeds or keep sheep best suited to their climate, with good mothering and milking abilities, based on the number or size of lambs they produce, overall health and hardiness, and more. Some can produce lambs "out of season" and not just in the spring, some breeds are better known for multiple births, and some resist foot rot or parasites better.

Generally speaking there are two types of sheep: *hair sheep* and *wool sheep*. Though the difference does not translate to their meat, hair sheep have begun to merit consideration from growers because they do not require shearing. They are also better suited to temperate climates and do not require their tails to be docked. Some popular hair breeds include the Blackbelly Barbados (reputed to have lower fat concentrations and milder meat among lambs), California Red, Dorper, and Katahdin. Dorpers are a meat breed out of South Africa that produces fast-growing and heavily muscled lambs. Katahdin is a hardy sheep good for forage.

Most sheep in the United States are Merino sheep, raised for both meat and wool. The Columbia sheep is an American breed developed by the United States Department of Agriculture, as is the Polypay. Hampshires are efficient and productive for both meat and wool in a variety of environments. The Texel is known for producing lean, heavily muscled lambs, while Suffolk is a fine all-purpose breed. The Cheviot is a small sheep that produces lambs with good dressing percentages. You might also recognize Romneys, Romanovs, or Rambouillets, which all sound like European dynasties to me.

What Cuts of Lamb Can I Order?

As with goat, it's nice to be able to make choices on a cut sheet knowing that you'll go through the meat in a relatively short time. I feel like I have to live with my lamb choices for less time than I do with beef or pork. The trade-off, I find, is that it's so easy to braise or roast big cuts of big animal with little tending, whereas I think more about my lamb meals and do more direct cooking (grilling, searing). As with goat, I absolutely love the versatility of ground lamb, and find myself grinding more on my own to use in sausages and for a range of Mediterranean dishes. Though I have used lamb and goat interchangeably in many dishes (and both in some, if I have them at the same time), I prefer the more tender texture of lamb for roasts, especially bone-in legs.

Primal Needs

The choices given on a cut sheet for lamb are relatively minimal; your choices are narrowed even further if they call you and offer instructions on the phone, as happened with both Puffy Legs and another lamb I ordered. In general, there are five primals to consider as you cut up your lamb. Even if the processor doesn't run through your choices by primal, you can do so as you make your decisions.

Though goat neck is often broken out as a distinct cut, especially because it's favored for curries and slow braises, I have never had lamb neck referenced specifically by processors, who will probably grind it up or cut it up as stew meat. Since I love ground lamb, that's never bothered me.

The **Shoulder** can be cut into stew or kabob meat, cut into shoulder chops, or made into a roast. Little shanky bits rest below, one of my favorite parts of a lamb.

Lamb Shoulder Cuts:

Stew meat	Blade chop
Ground meat	Shank
Shoulder or blade roast	

The **Rack** is the top middle of the lamb, but you already knew that, didn't you? You can have this cut as an entire rack roast, or have the rib chops cut individually. Treat these well—cook them with simple flavors and respect.

Lamb Rack Cuts:
Rack
Rib chop
Rib roast

The **Breast** rests below the rack and contains part of the forequarter and some ribs. I've gotten this as lamb riblets—a cut I had never encountered in a store. They have nice flavor, and I found them best roasted or grilled, though they are tiny little things, especially on a grass-fed lamb. If you want the breast for roasting or stuffing, tell your processor specifically.

Lamb Breast Cuts:
Breast
Riblets
Ground meat

The little **Loin** falls behind the rack and breast. Loin chops are tender and fabulous and easy to cook even if you forgot to defrost them overnight, and this is what you will generally receive if you don't specify that you'd like a roast instead. You can also ask to have double loin chops cut.

Lamb Loin Cuts:
Boneless loin roast
Loin chop
Double loin chop

Leg of lamb is such a satisfying thing to have on hand. I love whole legs so much that the only variation I usually make is to ask for one boned and tied and the other bone-in. You could get sirloin chops, though, if you're that kind of a person, or have the butcher cut or grind it up. But why not do that on your own and keep all possibilities alive? The rear shanks are substantial, full of flavor, and begging to be braised in whatever you like, from beer to wine to herbed stock.

Lamb Leg Cuts:
Whole leg, bone-in or boneless   Sirloin roast
Sirloin chop   Lamb shank

My lambs have always come packed with **Innards**. A little heart, some liver, and the kidneys are especially prized. Lamb tongue, like most tongue, is delicious, though lambs don't have very big tongues. I don't know what you might do with just one. Maybe serve it with champagne and eat like a supermodel.

How Much Will My Lamb Cost and How Much Will I Get?

Like goat, lamb makes a very nice whole-animal meat for an urban dweller. We are talking about a baby animal here, after all. Lambs deliver a nice mix of company cuts or those suitable for larger groups, like leg or a full rib rack, and smaller cuts like loin chops that are quick both to defrost and cook.

Maybe because you are dealing with a smaller animal, buying lambs is usually less complicated than buying a beef or a hog. Even if you are supposed to pay the producer one fee and the processor another, the producers usually break it out but have you pay the entire amount to them, including the cut-and-wrap and the "harvest" or kill fee.

Like other animals, in general you will pay per pound by hanging weight, though some producers sell whole or half-lambs at a set price. Most local grass-fed lambs I've found had hanging weights in the 50s, and rates that ranged from around $6 up to nearly $8 per pound, including all charges. Here are a few examples of how to calculate cost based on real examples of West Coast producers.

Producer Puffy Legs charges $6.50 per pound of hanging weight, plus $1.50 per pound for the harvest and cut-and-wrap. The lamb on offer has a hanging weight of 57 pounds:

$$57 \times (\$6.50 + \$1.50) = \$456$$

Gamboling Lamb Inc. sells grass-fed lamb for the inclusive rate of $6 per pound. So at a hanging weight of 52 pounds:

$$52 \times \$6 = \$312$$

I also found lambs sold for set prices by the half and whole. The grower should give you a range of either hanging weight or take-home weight that you will get for your price. Wooly Bully Farms, for example, sells whole lambs for $365 and half-lambs for $190, with 35 pounds and 17 pounds of take-home meat, respectively, including stew bones and some offal.

Questions to Ask Your Grower:

» Is the cost calculated by live weight or hanging weight, or is there a set cost?
» Are half-lambs available?
» How much deposit is required?
» Are your lambs fed grass or grain or both?
» What is the average hanging weight of your animals?
» Is the slaughter and processing included in the fee, or will I pay the processor separately?
» What time of year are your lambs available?

How Much Meat Will I Get?

Lambs usually come to market or are considered at market weight around 100 to 140 pounds, though this is variable. In general, grass-fed lambs weigh less than their grain-fed counterparts. The dressing percentage of a lamb is around 54 percent, which means that a lamb with a live weight of 125 pounds will deliver a hanging weight of about 68 pounds. You lose even more weight in the cut-and-wrap, of course, depending on how many bone-less cuts you ask for, the relative fattiness of your lamb, and other factors. Bone makes up about a fifth of the carcass, meat a little over half, and the rest is usually given over to fat. The grass-fed lambs I've ordered have delivered meat in the 30- to 40-pound range.

How Much Freezer Space Do I Need?

One whole lamb is just the right size to kind of screw you in the freezing department. You've got some big cuts in there, potentially, like a couple of meaty legs, plus another 25 pounds of meat. You might be able to fit it in your regular freezer compartment, but it will be completely full if you can. More likely, you'll go over. I'd arrange with a friend for surplus storage if you don't have a larger freezer on hand. If you order half a lamb, you're golden.

Who Will Kill My Lamb and How?

Your lamb will either be stunned or shot and then have its throat cut, or simply have its throat cut with a very sharp knife. Because lambs are small and relatively docile, many small producers kill and minimally process their own carcasses before transporting them to the processor for cut-and-wrap. Somehow that's how I think of lambs being killed, by someone in the Bible, with a beard. Others argue that killing animals simply by cutting their throats is potentially inhumane, as it takes a great deal of skill to do this perfectly and painlessly. Though I can't say that thinking of the slaughter is my favorite part of ordering an animal, the grower I talked to at length about it made me the happiest. He was thinking about it, now I was thinking about it, and I think we all benefitted.

Where Do I Pick Up My Meat?

In a strange turn of events, even though I paid only the producer for my lambs, I have always picked them up from the processor, though no money changed hands. Once, they had to call the grower to see if they could release my lamb to me, as my check had not yet reached him while I was eagerly awaiting my meat, cooler in hand. He let me take the lamb. So trusting.

Because lambs are small, you are more likely to be able to pick them up at the farmers market if your grower tables there. I have also found local growers who delivered to CSA drop locations.

When Will It Be Ready?

Many small growers raise sheep that only lamb once a year, which means you need to wait until they reach weight for yours to be slaughtered. For one of my lambs, I had to wait additional time for another order to come in, as that grower never has single animals harvested, preferring to send them in groups to reduce their stress. Once the animal is slaughtered, the meat is usually available in a matter of days.

RECIPES FOR LAMB

Lamb Kibbeh

Even though I'm sure I don't pronounce it correctly, I like to call these meaty ovals by their Turkish name, içli köfte. (Try dropping the name like a grandma from Istanbul, and hope none of your guests speaks Turkish.) I first made these as part of an appetizer array for an unconventional Thanksgiving. They were by far the favorite, and since they're football shaped, maybe they're more traditional than I gave them credit for. Truthfully, I had wanted to make kibbeh for years, intrigued by this traditional food served in countries including Israel, Syria, Lebanon, and Turkey. Nonvegetarian versions usually consist of fine bulgur wheat kneaded with ground meat and spices, then poached, boiled, baked, steamed, or fried. Some versions are even served raw. If you don't go crazy with the chile, children like them, too.

My delightful surprise in making them is that this is one of the most highly adaptable recipes around. Adjust the spices, the proportions, and the cooking method to suit. They are also delicious made with goat. Just don't call them "meatballs."

SERVES 6–8 AS AN APPETIZER

For the shell:

- 1½ cups bulgur wheat
- 1 small yellow onion, roughly chopped
- 2 teaspoons kosher salt
- 2 teaspoons Lam's Syrian Spice Blend (see the "Roast Chicken with Rice and Goat Stuffing" recipe in Chapter 8, "Goat")
- Pinch cayenne or Aleppo pepper, to taste
- A few grinds black pepper
- 1½ pounds ground lamb
- 1 ice cube

For the stuffing:

- 2 tablespoons olive oil
- ½ yellow onion, diced

1 teaspoon salt

½ teaspoon Lam's Syrian Spice Blend

½ teaspoon ground cinnamon

½ pound ground lamb

¼ cup parsley, roughly chopped, or mixed parsley and cilantro

¾ cup pine nuts, toasted

Cornstarch

Oil for frying

Yogurt

Fresh pomegranate seeds or pomegranate molasses (optional)

Butter lettuce leaves (optional)

Begin the shell by placing the bulgur in a bowl, and cover it with water. Let it sit for 10 minutes, then drain.

Pulse the onion in a food processor until finely chopped. Add the drained bulgur, the salt, and the spices, and process until the bulgur is finely ground. Add the lamb and one ice cube, and process until a smooth, stiff paste forms. Turn the paste out into a bowl, and cover. Refrigerate until very cold, 1 hour minimum.

In the meantime, make the stuffing. Heat the oil in a sauté pan over medium heat, and add the onion, salt, and spices. Cook until the onion is translucent and nearly golden. Add the lamb, and gently fry until cooked through. Add the pine nuts and parsley, and mix to incorporate. Turn the mixture out into a bowl to cool.

Lightly oil a sheet pan or cookie sheet. Put a pinch of salt and a spoonful of cornstarch in a bowl, and add a cup of cold water in a stream, whisking to dissolve the cornstarch and salt. With the stuffing and shell paste ready, wet your hands in the salted water, then pinch off a ball of shell paste slightly larger than a golf ball. Slightly cupping both your hands, slap the paste back and forth between them to form a roughly football-shaped lump.

With the lump of paste in the palm of your weak hand, insert the forefinger of your strong hand into a short end of the football. Keeping your finger inside the shell, rotate the shape, opening and closing your hand to keep the paste moving and make the walls smooth and even. (It's a strange movement to explain but very intuitive when you do it. Think Demi Moore in *Ghost* —I mean when she's making the pottery.) You should end up with a shape that looks like a tall vase about 3 inches long, with walls about ¼ inch thick, and an opening about 1 inch wide.

Insert about 1 tablespoon of filling into the opening. Wet your finger and dab it on the end of the kibbeh to seal and encase the filling. Using both palms, re-form the lump into a smooth football shape, and place on the oiled tray. Repeat with the remaining paste and filling, then cover them all with plastic wrap. (You can prepare the kibbeh up to this point an hour or two or even a day ahead, and keep refrigerated until you are ready to fry them. They also freeze well at this point.)

Heat ½ inch of oil in a large sauté pan over high heat. Place as many kibbeh in the pan as will fit without crowding, and fry, turning to crisp all sides, and lowering the heat if needed to ensure they don't brown too quickly. Drain the kibbeh briefly on paper towels as they come out of the oil. Continue cooking in batches.

There are as many ways to serve kibbeh as there are ways to cook them. One way I like is to mix plain Greek or strained yogurt with a little salt and a handful of pomegranate seeds and dollop it on the side. To make them into finger food, serve them Vietnamese spring-roll style: Place one kibbeh on a leaf of butter lettuce, dollop a little yogurt and add a tiny drizzle of pomegranate molasses if you like, then wrap and eat.

The Kibbeh Finger

○ ○ ○ ○ ○ ○ ○ ○　○ ○ ○ ○ ○ ○ ○　○ ○ ○ ○ ○ ○ ○

Paula Wolfert's *The Cooking of the Eastern Mediterranean* has a characteristically lovely note about women known for their "kibbeh fingers," and I must admit that the description of these gnarled women throwing down with the kibbeh production was part of the intrigue in wanting to make them. Dexterity is useful indeed, along with a long, elegant forefinger to help with shaping the shell. But if you have hands like oven mitts, your kibbeh will still be delicious, even if they're shaped into patties or balls. So don't let lack of a kibbeh finger hold you back. Alternatively, scan the digits of your friends and target one with nice piano hands. Trade a package of ground lamb for her kibbeh shaping, while you sit back with your huge mitts wrapped around a glass of wine.

Lamb Tagine with Pears and Sour Cherries

This is a wonderful braise for the weekend or a work-from-home day when you can leave the pot on the stove for a few hours. I made this with lamb shanks, but shoulder or neck would work, too—any cut that can take the time on the stove. Make sure you buy or harvest firm pears; if they're too ripe they'll turn to mush in the tagine. I used a bottle of Moscato d'Asti I had on hand, but sherry would work just fine. Or just use water.

SERVES 4–6

2 tablespoons olive oil

4 lamb shanks, or one lamb shoulder roast

4 large shallots, sliced

1 small yellow onion, chopped

1 teaspoon coriander seeds, ground

Pinch ground cardamom

2 cinnamon sticks

½ cup Moscato or sherry, plus a splash more

1 teaspoon saffron threads

⅓ cup dried sour cherries

3 tablespoons butter

3 unpeeled Bosc pears, cored and cut into sixths

2 tablespoons each chopped parsley and cilantro

Heat the olive oil over medium-high heat in a Dutch oven or other heavy pot with a lid. Season the shanks or shoulder with plenty of salt and pepper, add them to the pot, and brown on all sides. Add the shallots and onion along with the ground coriander, cardamom, and cinnamon sticks, and cook, stirring, until the vegetables are softened.

Deglaze the pan with Moscato or sherry. Add enough water to come halfway up the meat, and cover the pan. Reduce the heat to low, and cook at a bare simmer until meat is falling off the bone, at least 2 hours. (Alternatively, cook in a 300°F oven instead of on the stove.)

Uncover the pot. Mix the saffron with a little of the braising liquid, and allow it to bloom. Add to the pot along with the cherries. Simmer until the liquid is reduced and thickened, about 15 minutes.

While the liquid reduces, heat the butter in a sauté pan over medium heat. Add the pear slices and cook until caramelized on all sides and softened. Deglaze the pan with another splash of wine if you like. Add the pears to the tagine, and gently mix to coat the pears and heat through. Sprinkle the tagine with parsley and cilantro and serve.

Middle Eastern Skewers

This is less of a recipe than an idea, but it's a fine idea and one that I employ with abandon, especially in summer, when it's easier to grill outside and not heat up your whole house. The basis of the idea is onion juice, which Middle Eastern cooks know both tenderizes and flavors all sorts of meat, rendering even tougher cuts tender. Swish the lamb or goat chops in the marinade for only a half-hour before grilling or searing. Throw in some herbs and any other bright flavors you want, let the meat take a bath in that for a while, then grill and serve. Flatbread is nice. Yogurt is too, on the side, with a little salt and crushed garlic, unless you add yogurt to the marinade. Sliced cucumbers, sliced tomatoes, even an orange and olive salad with parsley, would be lovely on the side.

SERVES 6

1 medium yellow onion, peeled
2 garlic cloves
Handful parsley stems and leaves
Handful cilantro stems and leaves
Juice of 1 lemon
3 tablespoons red wine vinegar
⅓ cup good olive oil
1½ pounds cubed lamb or goat leg, or beef cut from the chuck, trimmed of gristle and sinew

Quarter the onion, and place it in a food processor with the garlic, parsley, and cilantro. Add 2 pinches of salt, and pulse to chop the herbs. Add the lemon juice and vinegar, and process on high until you get a loose slurry. With the food processor running, add the olive oil in a stream.

Place the marinade and meat in a resealable bag. Squish around to coat, and let hang out in the fridge or on the counter if cool for 1 to 3 hours (overnight will do some serious tenderizing bordering on mushy, so don't do it unless you mean it).

Allow the meat to come to room temperature before grilling. Remove from the marinade and wipe off any extra marinade with your finger. Heat a grill or grill pan on high. Skewer the meat, allowing plenty of room between cubes for even searing. Grill, turning frequently, for about 8–10 minutes, or until pink inside but well marked on the exterior. Serve.

Any leftovers are splendid tucked into a Greek pita with veg for lunch the next day.

Indian variation: Omit the lemon juice from the marinade, and add 1 cup plain yogurt and 1 heaping tablespoon garam masala plus 1 teaspoon cracked coriander seeds to the marinade. Proceed.

Lamb with Honey

This has been our go-to lamb stew for years, a Moorish–Spanish pot of fun that children eat with nary a bit of cajoling. Though I specify lamb here, the flavors would go nicely with pork—fresh leg would be nice. It may seem counterintuitive to add the shredded lemon, but it acts much the way preserved lemon might, and you're probably more likely to have a fresh lemon on hand.

SERVES 4–6

½ teaspoon saffron threads

1 slice organic or unsprayed lemon, seeds removed

1 boneless leg of lamb, about 4 pounds

¼ cup olive oil

1 large onion, cut in half lengthwise and then thinly sliced

2 fresh poblano chiles, stemmed and seeded, membrane removed and flesh slivered

2 large garlic cloves, peeled and thinly sliced

½ cup dry white wine

3 tablespoons brandy

1 teaspoon ground cinnamon

1 tablespoon smoked paprika

⅓ cup honey

3 tablespoons sherry vinegar

Steep the saffron in a little warm water. Shred the lemon, peel and all, with a chef's knife and reserve it.

Cut the lamb into 1- by 2-inch pieces, and season well with salt and pepper. Heat the oil in a Dutch oven or heavy pot over medium-high heat. Add the lamb and brown it well, then remove from the pot. Add the onion, chiles, and garlic to the pot and cook, scraping up bits from the bottom, until soft.

Deglaze the pot with the wine and brandy, stirring well. Add the cinnamon, saffron in liquid, lemon, smoked paprika, and the lamb, stirring to coat. Cover and cook for 45 minutes to 1 hour, or until the meat is tender.

Just before serving, add the honey and vinegar and stir well. Cook for 5 more minutes to blend the flavors.

Irish Death Lamb Stew

It's so nice when eating local includes both pastured lamb and stout brewed just down the road. This is Jeff's creation, a stew enhanced by a beer from Iron Horse Brewery in Ellensburg, Washington, and given extra depth by roasting the bones to make the stock.

Though Jeff uses a regular roux to thicken the stew, I find that using a little something those sexy French call beurre manié gives it a nice gloss and body. It's simply flour and butter kneaded together to form a paste, then whisked into the stew.

SERVES 6–8

2 pounds bone-in lamb shoulder steaks, meat cut off the bone and bones reserved

1 pound lamb neck slices, meat cut off the bone and bones reserved

2 cups rich chicken stock

2 22-ounce bottles Iron Horse Irish Death or other stout

Olive oil for frying

1 cup diced bacon ends

5 tablespoons flour

1 yellow onion, cut in small dice

2 ribs celery, diced

1 tablespoon fresh thyme leaves

1 tablespoon freshly cracked black pepper

1 tablespoon anchovy paste

1 tablespoon soy sauce

3 medium Yukon gold potatoes, cubed

3 medium carrots, cut into chunks

2 cups beef stock

2 tablespoons softened butter

½ teaspoon wine vinegar

½ cup minced parsley

Preheat the broiler.

Place the lamb bones on a foil-lined baking sheet. Roast them under the broiler until well browned, then place them in a small stockpot. Add the chicken stock and one bottle of stout. Simmer over low heat while you get the rest of the stew together.

Preheat the oven to 350°F.

Place 1 tablespoon of olive oil in a large Dutch oven, and add the bacon. Sauté until the bacon has released its fat and is browned and crisp. Remove from the pan and set it aside.

Brown the lamb in the bacon fat. Add 3 tablespoons of the flour and stir to coat, then cook for 3 minutes. Add the onions, celery, thyme, pepper, anchovy paste, and soy sauce, and cook over medium heat until the onion and celery are soft. Add the potatoes, carrots, beef stock, and lamb–stout stock, along with half of the remaining bottle of stout.

Bring the pot to a boil, stir, and cover. Place in the oven, and cook for 1¾–2 hours, or until the lamb is tender and the sauce has body. Remove the pot from the oven, uncover, and place on the stove over low heat. Knead together the butter and remaining flour to form a paste, then whisk to incorporate it into the stew. Simmer the stew on low for an additional 10 minutes or until thickened.

Just before serving, add the vinegar, stir, and taste for salt. Garnish with the parsley and serve.

Lakhchak

Ashak is the more famous cousin of lakhchak, but it requires assembling individual dumplings, and while I like to think I'm someone who will do that on a weeknight, it's just not true. Lakhchak is a much simpler dish to prepare, and even if calling this Afghani lasagna is overly simplistic, if it gets you to try it then why not. Layers of pasta and lamb are sauced with yogurt, not cheese, making for a subtly spiced, lighter dish. This is a favorite of my dear Portland colleague Christina and her food-adventurer hubbie, Rick. I'm indebted to them for the brilliant inspiration and special touches, like the fried leeks (which add flavor and a welcome spot of color) and the shower of fresh mint on top. Use thick Greek yogurt or strain a good-quality brand like Strauss to avoid a watery sauce.

SERVES 4

⅓ cup plus 2 tablespoons olive oil

2 medium yellow onions, chopped

1 pound ground lamb, goat, or grass-fed beef

½ teaspoon ground cinnamon

1 teaspoon grated ginger

1½ teaspoons coriander seeds, toasted and ground to a powder

½ cup tomato juice

2 large leeks, white parts only

1 garlic clove, peeled

2 cups whole fat Greek or strained yogurt

1 pound fresh pasta sheets

6 fresh mint leaves, rolled and sliced in a chiffonade

To make the meat sauce, heat ⅓ cup of oil in a sauté pan over medium heat, and add the onions. Fry, stirring constantly, until lightly browned. Add the meat and brown it. Add the cinnamon, ginger, and coriander, then deglaze the pan with the tomato juice and bring to a boil. Reduce the heat and simmer for a half-hour or until thick.

While the meat sauce is cooking, thinly slice the leeks and rinse and dry carefully. Heat the remaining olive oil in a clean sauté pan, and add the leeks. Cook, stirring frequently, until limp and nearly golden.

Run the garlic clove over a microplane into the yogurt, season with a bit of salt, and set aside at room temperature.

Bring a pot of water to a boil, and salt it. Add the pasta, cook until al dente, and drain. Spread out about one third of the yogurt in a warm, shallow serving dish. Top with half of the cooked pasta, then another third of the yogurt. Sprinkle half of the leeks evenly over the top, and then half of the meat sauce. Top with the remaining pasta, then layer the remaining yogurt, meat sauce, and leeks. Sprinkle with the mint and serve immediately.

Lamb Shank Ragu over Polenta

Though you might think of this as winter food, I made it once for the whole crowd when my family went to the Oregon coast in August. By the time the evening wind was whistling over the sand and it was chilly enough for sweaters, the shanks were melting into their braising liquid and fighting the sea to fill the air with their delicious aroma. Atticus, who like other fussy meat eaters blanches when confronted with cartilage and bone, inhales this when the meat is shredded in the kitchen. That also makes it easy to pool on top of the polenta and serve family-style.

SERVES 8

8 grass-fed lamb or goat shanks

⅓ cup olive oil

2 medium yellow onions, chopped

4 carrots, peeled and diced

3 large garlic cloves, minced

1-inch-wide swatch unsprayed orange rind, removed with a
    vegetable peeler

6 sprigs fresh thyme

2 large sprigs rosemary

½ 28-ounce can San Marzano tomatoes, drained and crushed

1 bottle pinot noir wine

2 cups fine polenta

2 cups finely grated Parmesan

4 tablespoons soft butter

Season the shanks with plenty of salt and freshly ground pepper. Heat the olive oil over medium-high heat in a large Dutch oven. When it's hot, add the shanks, and cook, turning frequently, until browned, working in batches if necessary to get a good sear. Remove the shanks to a plate.

Preheat the oven to 300°F.

Lower heat under the pot to medium-low, and add the onions, carrots, garlic, orange peel, and herb sprigs to the pot, scraping up the bits from the bottom of the pan. Cook until the onion is golden. Add the tomatoes, and stir to combine. Nestle the shanks in among the vegetables, and add enough wine to come halfway up the shanks. Bring to a simmer, cover, then place the pot in the oven. Braise for about 3 hours, or until the meat is falling off the bone.

Remove the shanks, and shred the meat. Remove the herb sprigs and orange peel, and reduce the liquid slightly. Add enough liquid to form a rich sauce.

Bring 8 cups of water to a boil in a large saucepan, and add a pinch of salt. Whisk the polenta into the water in a steady stream, whisking constantly to avoid lumps. Reduce the heat to low and cook, stirring to ensure that the polenta doesn't stick to the bottom. Cook for 25–30 minutes, or until the polenta is creamy. Add the cheese, and taste for salt. When ready to serve, stir in the soft butter and turn the polenta into a large, warm serving bowl. Top with the ragu.

Red-Cooked Lamb

Originally I used this recipe to cook pork belly, and I won't argue with you if that's the direction you want to go. But after reading about using the same Chinese technique for lamb, I was intrigued. The spices and the long braise work incredibly well with the deeper flavors of lamb, especially with richer cuts like shoulder. Chinese rock candy is also called rock sugar, and it's readily available in Asian groceries. Double-bag the sugar, and hit it with a meat mallet or hammer to crush the crystals, so that it dissolves more readily. This is especially good served with Vietnamese broken rice, a favorite in our house, but plain sticky rice will do.

SERVES 4

½ cup Shaoxing wine or dry sherry
2 tablespoons soy sauce
Rough ½ cup crushed Chinese rock candy
1½ pounds boneless lamb shoulder, cubed
2 whole star anise
1 cinnamon stick
1 teaspoon Szechuan peppercorns (optional)
1 teaspoon fennel seeds
1 teaspoon coriander seed
2 garlic cloves, peeled and smashed
1-inch-thick piece fresh ginger, well smashed with a chef's knife

Combine 6 cups of water, the wine, soy sauce, and the rock sugar in a Dutch oven, and bring to a boil. Add the lamb, and lower the heat to medium-high to maintain a brisk simmer. Cook for 15 minutes, skimming frequently. Add the star anise, cinnamon, peppercorns, fennel, coriander, garlic, and ginger. Lower the heat to a bare simmer and cook, partially covered, for 2–3 hours, or until the lamb is quite tender.

Remove the lamb from the pot, and keep covered. Pour the liquid through a fine-mesh sieve into a medium saucepan. Bring to a boil over high heat, and reduce until slightly thickened. Moisten the lamb with thickened juices to taste, and serve.

ACKNOWLEDGMENTS

Uncle Dave's Cow would not be possible were it not for Uncle Dave, of course, and Cousin Kate, and Jason, and Aunt Marilyn, who all helped so much on this book, as well as the rest of the farming Fenn community of uncles, aunts, and cousins who shaped my life, especially the matriarch, Grams.

The entire Cast of Characters was and is invaluable, but my own dear nuclear family supported so kindly a wife and mother who was carcass-obsessed as well as locked away in her study for long periods. Husband Hendel, Atticus, and Mo, a million thanks for your love and support, and for eating so much pot roast and too-spicy sausage.

The Girls Friday encouraged me to do the project, read first drafts, collaborated on recipes, and weighed in on cover design. They made me sign up for Twitter and kept me company when I went to fetch pigs. They do it all. XOXO to Flomar, Chenry, Freedom, and Tito.

I both thank and blame my family of origin for the peculiar direction my culinary life has taken; I have to believe that flesh production and sausage making are in my DNA. Thanks to my parents, for planting those locker meat memories, and to my dear brothers, Dee, Erik, and Jeff, for remaining my very favorite cooking partners.

In a publishing world that can be inhospitable to the niche-of-a-niche project, Kate Rogers was the first to love and believe in the holistic idea behind *Uncle Dave's Cow*, and an unflagging supporter even when my meat journeys took me to unexpected places. Janet Kimball and Margaret Sullivan helped make a mere manuscript into a book.

Ayun Halliday—Milky can't thank you enough for capturing absolutely perfectly the spirit of this book and the characters that inhabit its pages, including those who are no longer with us. Uncle Dave's Cow, Shorthorn, Puffy Legs, and Handsome Trotters would all like to say, you did them justice.

And to all the farmers, big and small, who take care and pride in the animals they raise to be food and have patience with urban eaters on a steep learning curve, thank you for doing an incredibly hard and really important job.

GLOSSARY

Captive-bolt stunner: A device consisting of a steel bolt propelled by compressed air through a metal sleeve. Used to render an animal unconscious prior to slaughter.

Carcass cutting yield: The expressed proportion of **cutout weight** to **hanging weight**.

Charcuterie: Cured, salted, smoked, or preserved meats, or meat products such as pâtés, sausages, terrines, or hams; can also refer to a store that sells these products.

Cold smoking: A technique wherein foods are flavored, but not cooked, by placing them in an enclosed chamber with a smoke source—either an external smoke chamber or a wood-burning fire placed at a distance from the food to be smoked. Meats that are cold-smoked are usually salted or brined first to inhibit the growth of bacteria, and usually cooked after smoking.

Community Supported Agriculture (CSA): An arrangement whereby consumers buy subscriptions to locally grown produce or meat and in return receive regular boxes or bundles of food from the farm throughout a specified season.

Cut sheet: Instructions given to the processor that specify how the carcass should be cut and the weights of meat or number of cuts to include per package, as well as any additional details about processing such as smoking or making sausage.

Cut-and-wrap: The butchering of a carcass into retail cuts, and the packaging of those cuts in butcher paper or vacuum-sealed plastic.

Cutout weight (also **finished cut weight** or **cutout**): The weight of take-home product cut from a carcass.

Dressing percentage: The expressed proportion of **hanging weight** or **hot carcass weight** to the **live weight** of the animal.

Dry aging: A process in which meat, typically beef, is hung in a temperature- and humidity-controlled environment for a period of days or weeks. During this time, enzymes break down muscle fibers in the meat and moisture evaporates, resulting in a more tender product with concentrated flavor.

Grower: A farmer who raises animals for meat.

Hanging weight (also **hot carcass weight** or **HCW**): The weight of the unchilled carcass—usually with the blood, viscera, head, hide, and feet removed—that usually forms the basis for cost when purchasing a whole or partial animal.

Hot smoking: A technique wherein foods are both cooked and flavored by placing them in a heated chamber with burning wood, wood chips, or compressed wood pieces.

Kill fee: Money paid for the slaughter of an animal separate from the cost of the meat.

Live weight (also **on the hoof**): The weight of a live animal.

Mobile slaughter unit (MSU) (also **mobile processing**): Specially built trucks equipped with generators, hot and cold water, refrigeration, and processing tools that travel directly to farms to slaughter and process animal carcasses onsite.

Offal (also **variety meats or organ meats**): The organs and extremities of an animal consumed by people. Includes entrails such as heart, liver, and kidney, as well as tongue, head, foot, cheek, and tail.

Primal (also **wholesale cut**): A large subsection of a carcass that is further divided into retail cuts.

Processor: A person or company that slaughters and butchers an animal.

Side: A half-carcass that includes the hind- and forequarters.

Sticking: Cutting the jugular vein and carotid artery of an animal during butchering, in order to drain the blood from the body.

SELECTED RESOURCES

Grass-Fed Nutrition Information and Resources

Pastured meat and dairy information, including a state-by-state list of farms raising pastured meat, is available at **eatwild.com**.

Nutrition Journal has published several articles on the relative fat content of grass- and grain-fed meat. See "A review of fatty acid profiles and antioxidant content in grass-fed and grain-fed beef," by Cynthia A. Daley et al., in the 10 March 2010 issue, available online at **nutritionj.com**.

The American Grassfed Association, an advocacy group for growers of grass-fed meat, can be found at **americangrassfed.org**.

A review of the USDA's food labeling terms, including grass-fed marketing claim standards, can be found at the Agricultural Marketing Service site at **ams.usda.gov**.

The *Journal of Animal Science* has many excellent articles on the effects of various feed types in animals. Visit **jas.fass.org**.

Humane Slaughter

Dr. Temple Grandin is the guru of humane livestock handling and humane slaughter. Visit **grandin.com** to learn more.

A short overview of humane livestock handling and slaughter guidelines is available at the Food Safety and Inspection Service website, **fsis.usda.gov**.

A complete overview of the Humane Methods of Slaughter Act is available at **awic.nal.usda.gov/government-and-professional-resources/federal-laws /humane-methods-slaughter-act**.

Local Food, Food Hubs, and Regional Food Infrastructure

For a national directory of farmers markets, community supported agriculture (CSA) programs, and local food resources, visit **localharvest.org**.

Washington State University Extension offers a detailed resource site on buying local meat, at **extension.org/pages/18168/local-meat-buying-guide**. The site includes a free download of the "Beef and Pork Whole Animal Buying Guide," a supremely helpful text (with pictures) suited for beginners, from Iowa State University Extension. In general, university extension programs are an excellent source of information and resources for meat butchery and buying guides.

If you live in the Western United States, visit **food-hub.org**.

For national food hub resources, visit the USDA site at **ams.usda.gov/AMS v1.0/foodhubs**.

Cookbooks and Resource Texts

Charcuterie: The Craft of Salting, Smoking, and Curing by Michael Ruhlman and Brian Polcyn (W. W. Norton & Company, 2005). An excellent resource text for amateur and advanced smokers and sausage makers.

Cured: Slow Techniques for Flavoring Meat, Fish and Vegetables by Lindy Wildsmith (Krause, 2010). A beautiful book with simple techniques from potting meat to curing salmon.

Goat: Meat, Milk, Cheese by Bruce Weinstein and Mark Scarbrough (Stewart, Tabori & Chang, 2011). Hysterical prose and great recipes for that most underappreciated meat.

Good Meat: The Complete Guide to Sourcing and Cooking Sustainable Meat by Deborah Krasner (Stewart, Tabori & Chang, 2010). A lovely resource book on pastured meat and poultry.

Great Sausage Recipes and Meat Curing by Rytek Kutas (The Sausage Maker Inc., 1984). My oldest brother would tell you this is the only book you need for curing meat.

Mediterranean Grains and Greens: A Book of Savory, Sun-Drenched Recipes by Paula Wolfert (Ecco, 1998). For those who want a helping of vegetables with their meat. Oh heck, any of Wolfert's books are amazing (paula -wolfert.com).

Plenty: Vibrant Recipes from London's Ottolenghi by Yotam Ottolenghi (Chronicle, 2011). When you need a break from meat, this is one of the most glorious meatless cookbooks ever.

The Art of Charcuterie by The Culinary Institute of America (Wiley, 2010). For those who wish to make twenty pounds of emulsion sausages at a time, this is the book for you. Excellent explanations of the science of curing.

The Complete Book of Butchering, Smoking, Curing, and Sausage Making: How to Harvest Your Livestock & Wild Game by Philip Hasheider (Voyageur, 2010). A graphic (in all senses of that word) how-to book for those who want to butcher and dress their kill.

The Whole Beast: Nose to Tail Eating by Fergus Henderson (Ecco, 2004). Deservedly a classic of British cookery.

INDEX